# SHELSEAS MAGIC SQUARE II
## It Works

Magic Square is a mathematical formula containing rows, columns and diagonals with the same sum of integer as an entire whole. Over the years, around the world, people have enjoyed this magnificent symmetry of numbers.

Add each square of numbers in any row, column, or diagonal for the same sum of 260

| 64 | 2  | 3  | 61 | 60 | 6  | 7  | 57 |
|----|----|----|----|----|----|----|----|
| 9  | 55 | 54 | 12 | 13 | 51 | 50 | 16 |
| 17 | 47 | 19 | 45 | 44 | 22 | 42 | 24 |
| 40 | 26 | 38 | 28 | 29 | 35 | 31 | 33 |
| 32 | 34 | 30 | 36 | 37 | 27 | 39 | 25 |
| 41 | 23 | 43 | 21 | 20 | 46 | 18 | 48 |
| 49 | 15 | 14 | 52 | 53 | 11 | 10 | 56 |
| 8  | 58 | 59 | 5  | 4  | 62 | 63 | 1  |

# Magic Squares

The structure of the number fill-ins [A thru D] will show you a creative method to sharpen your skills at counting & calculations to proof for a solved Magic Square.

# SHELSEAS MAGIC SQUARE II
## It Works

### Positive [C]

|   | 2 | 3 |   |   | 6 | 7 |   |
|---|---|---|---|---|---|---|---|
| 9 |   |   | 12 | 13 |   |   | 16 |
| 17 |   | 19 |   |   | 22 |   | 24 |
|   | 26 |   | 28 | 29 |   | 31 |   |
|   | 34 |   | 36 | 37 |   | 39 |   |
| 41 |   | 43 |   |   | 46 |   | 48 |
| 49 |   |   | 52 | 53 |   |   | 56 |
|   | 58 | 59 |   |   | 62 | 63 |   |

### Negative [C]

| 64 |   |   | 61 | 60 |   |   | 57 |
|---|---|---|---|---|---|---|---|
|   | 55 | 54 |   |   | 51 | 50 |   |
|   | 47 |   | 45 | 44 |   | 42 |   |
| 40 |   | 38 |   |   | 35 |   | 33 |
| 32 |   | 30 |   |   | 27 |   | 25 |
|   | 23 |   | 21 | 20 |   | 18 |   |
|   | 15 | 14 |   |   | 11 | 10 |   |
| 8 |   |   | 5 | 4 |   |   | 1 |

Each mathematical formula will introduce unique geometric patterns of shaded squares through the symmetric two parts in each Magic Square.

# SHELSEAS MAGIC SQUARE II
## It Works

### Description

It Works will offer [20] Magic Squares

Integer:  In each Magic Square the integer as an entire whole is symbolic to each mathematical formula.

Double Even:  Even number of rows & columns, divide by two, for even number of shaded squares in each row & column.

Symmetric:  Two parts of Positive & Negative in each Magic Square.

Geometric:  Unique patterns of shaded squares in each Magic Square.

Guideline:  The mathematical formula $4 \times 4 + 1 \times 2 = 34$ will be shown for a solved Magic Square in the number fill-ins [A thru D] format.

Extension:  Each mathematical formula will be shown as an extension in the number fill-ins [A thru D] in each Magic Square.

Objective:  Combine both Positive & Negative number fill-ins [A thru D] in each Magic Square.  Add each square of numbers in any row, column, or diagonal to proof the same sum of integer for a solved Magic Square.

# "Welcome to my world of Magic Squares"

Created by,
Peggy Brown

SHELSEAS MAGIC SQUARE II
It Works

## Guideline

# A

Positive [A] Start with the top row, left to right in each row, going down. Begin with the number shown hint, count each square, only fill-in shaded squares.

Negative [A] Start with the bottom row, right to left in each row, going up. Begin with the number shown hint, count each square, only fill-in shaded squares.

# B

Positive [B] Start with the top row, right to left in each row, going down. Begin with the number shown hint, count each square, only fill-in shaded squares.

Negative [B] Start with the bottom row, left to right in each row, going up. Begin with the number shown hint, count each square, only fill-in shaded squares.

# C

Positive [C] Start with the bottom row, right to left in each row, going up. Begin with the number shown hint, count each square, only fill-in shaded squares.

Negative [C] Start with the top row, left to right in each row, going down. Begin with the number shown hint, count each square, only fill-in shaded squares.

# D

Positive [D] Start with the bottom row, left to right in each row, going up. Begin with the number shown hint, count each square, only fill-in shaded squares.

Negative [D] Start with the top row, right to left in each row, going down. Begin with the number shown hint, count each square, only fill-in shaded squares.

The Number Fill-ins [A] Will Be Your Guideline
For A Solved Double Even – Magic Square.

# Number
# Fill-ins

# [A]

Positive [A] Start with the top row, left to right in each row, going down.  Begin with the number shown hint, count each square, only fill-in shaded squares.

Negative [A] Start with the bottom row, right to left in each row, going up.  Begin with the number shown hint, count each square, only fill-in shaded squares.

Objective:  Combine both Positive & Negative number fill-ins [A] in each Magic Square.  Add each square of numbers in any row, column, or diagonal to proof the same sum of integer for a solved Magic Square.

## Mathematical Formulas

$$8 \times 8 + 1 \times 4 = 260$$

$$12 \times 12 + 1 \times 6 = 870$$

$$16 \times 16 + 1 \times 8 = 2056$$

$$20 \times 20 + 1 \times 10 = 4010$$

$$24 \times 24 + 1 \times 12 = 6924$$

## SHELSEAS MAGIC SQUARE II
### It Works

Positive [A] Start with the top row, left to right in each row, going down.  Begin with the number 1, count each square, only fill-in shaded squares.

| 1 |   |   | 4 |
|---|---|---|---|
|   | 6 | 7 |   |
|   | 10 | 11 |   |
| 13 |   |   | 16 |

Negative [A] Start with the bottom row, right to left in each row, going up.  Begin with the number 2, count each square, only fill-in shaded squares.

|   | 15 | 14 |   |
|---|---|---|---|
| 12 |   |   | 9 |
| 8 |   |   | 5 |
|   | 3 | 2 |   |

Combination of both Positive & Negative number fill-ins [A] is shown for a solved Magic Square.  Objective:  Add each square of numbers in any row, column, or diagonal to proof the same sum of 34.

| 1 | 15 | 14 | 4 |
|---|---|---|---|
| 12 | 6 | 7 | 9 |
| 8 | 10 | 11 | 5 |
| 13 | 3 | 2 | 16 |

Magic Square
Number Fill-ins [A]
Mathematical Formula 4 X 4 + 1 X 2 = 34

# SHELSEAS MAGIC SQUARE II
## It Works

Positive [A] Start with the top row, left to right in each row, going down.  Begin with the number 2, count each square, only fill-in shaded squares.  Helpful hints below.

|   | 2 |   |   |   |   | 7 |    |
|---|---|---|---|---|---|---|----|
| 9 |   |   |   |   |   |   | 16 |
|   |   | 19 |   |   | 22 |   |    |
|   |   |   | 28 | 29 |   |   |    |
|   |   |   | 36 | 37 |   |   |    |
|   |   | 43 |   |   | 46 |   |    |
| 49 |   |   |   |   |   |   | 56 |
|   | 58 |   |   |   |   | 63 |   |

Negative [A] Start with the bottom row, right to left in each row, going up.  Begin with the number 1, count each square, only fill-in shaded squares.  Helpful hints below.

| 64 |   |   |   |   |   |   | 57 |
|----|---|---|---|---|---|---|----|
|    | 55 |   |   |   |   | 50 |   |
|    |   | 45 | 44 |   |   |   |   |
|    |   | 38 |   |   | 35 |   |   |
|    |   | 30 |   |   | 27 |   |   |
|    |   |   | 21 | 20 |   |   |   |
|    | 15 |   |   |   |   | 10 |   |
| 8  |   |   |   |   |   |   | 1  |

Objective:  Combine both Positive & Negative number fill-ins [A].  Add each square of numbers in any row, column, or diagonal to proof the same sum of 260 for a solved Magic Square.  Helpful hints below.

| 64 | 2  |    |    |    |    | 7  | 57 |
|----|----|----|----|----|----|----|----|
| 9  | 55 |    |    |    |    | 50 | 16 |
|    |    | 19 | 45 | 44 | 22 |    |    |
|    |    | 38 | 28 | 29 | 35 |    |    |
|    |    | 30 | 36 | 37 | 27 |    |    |
|    |    | 43 | 21 | 20 | 46 |    |    |
| 49 | 15 |    |    |    |    | 10 | 56 |
| 8  | 58 |    |    |    |    | 63 | 1  |

Magic Square – Number Fill-ins [A] – Mathematical Formula 8 X 8 + 1 X 4 = 260

# SHELSEAS MAGIC SQUARE II
## It Works

Positive [A] Start with the top row, left to right in each row, going down.  Begin with the number 1 count each square, only fill-in shaded square.  Helpful hints below.

| | | | | | | | | | | | |
|---|---|---|---|---|---|---|---|---|---|---|---|
| 1 | | | | | | | | | | | 12 |
| | 14 | | | | | | | | | 23 | |
| | | | 29 | | | 32 | | | | | |
| | | 40 | | | | | | 45 | | | |
| | 51 | | | | | | | | 58 | | |
| | | | | | 66 | 67 | | | | | |
| | | | | | 78 | 79 | | | | | |
| | 87 | | | | | | | | 94 | | |
| | | 100 | | | | | | 105 | | | |
| | | | 113 | | | 116 | | | | | |
| | 122 | | | | | | | | | 131 | |
| 133 | | | | | | | | | | | 144 |

Negative [A] Start with the bottom row, right to left in each row, going up.  Begin with the number 4, count each square, only fill-in shaded squares.  Helpful hints below.

| | | | | | | | | | | | |
|---|---|---|---|---|---|---|---|---|---|---|---|
| | | | 141 | | | | | | 136 | | |
| | | | | 128 | | | 125 | | | | |
| | | | | | 115 | 114 | | | | | |
| 108 | | | | | | | | | | | 97 |
| | 95 | | | | | | | | | 86 | |
| | | 82 | | | | | | | 75 | | |
| | | 70 | | | | | | | 63 | | |
| | 59 | | | | | | | | | 50 | |
| 48 | | | | | | | | | | | 37 |
| | | | | | 31 | 30 | | | | | |
| | | | | 20 | | | 17 | | | | |
| | | | | 9 | | | | 4 | | | |

Magic Square
Number Fill-ins [A]
Mathematical Formula 12 X 12 + 1 X 6 = 870

# SHELSEAS MAGIC SQUARE II
## It Works

Positive [A] Start with the top row, left to right in each row, going down.  Begin with the number 1, count each square, only fill-in same shaded squares.

Negative [A] Start with the bottom row, right to left in each row, going up.  Begin with the number 4, count each square, only fill-in same shaded squares.

Objective:  Combine both Positive & Negative number fill-ins [A].  Add each square of numbers in any row, column, or diagonal to proof the same sum of 870 for a solved Magic Square.  Helpful hints below.

| 1 |  |  | 141 |  |  |  |  | 136 |  |  | 12 |
|---|---|---|---|---|---|---|---|---|---|---|---|
|  | 14 |  |  | 128 |  | 125 |  |  |  | 23 |  |
|  |  |  | 29 | 115 | 114 | 32 |  |  |  |  |  |
| 108 |  |  | 40 |  |  |  |  | 45 |  |  | 97 |
|  | 95 | 51 |  |  |  |  |  |  | 58 | 86 |  |
|  |  | 82 |  |  | 66 | 67 |  |  | 75 |  |  |
|  |  | 70 |  |  | 78 | 79 |  |  | 63 |  |  |
|  | 59 | 87 |  |  |  |  |  |  | 94 | 50 |  |
| 48 |  |  | 100 |  |  |  |  | 105 |  |  | 37 |
|  |  |  |  | 113 | 31 | 30 | 116 |  |  |  |  |
|  | 122 |  |  | 20 |  |  | 17 |  |  | 131 |  |
| 133 |  |  | 9 |  |  |  |  | 4 |  |  | 144 |

Magic Square
Number Fill-ins [A]
Mathematical Formula 12 X 12 + 1 X 6 = 870

# SHELSEAS MAGIC SQUARE II
## It Works

Positive [A] Start with the top row, left to right in each row, going down. Begin with the number 1, count each square, only fill-in shaded squares. Diagonals are shown for the same sum of 2056. Helpful hints below.

| 1 | | | | | | | | | | | | | | | 16 |
|---|---|---|---|---|---|---|---|---|---|---|---|---|---|---|---|
| | 18 | | | | | | | | | | | | | 31 | |
| | | 35 | | | | | | | | | | | 46 | | |
| | | | 52 | | | | | | | | | 61 | | | |
| | | | | 69 | | | | | | | 76 | | | | |
| | | | | | 86 | | | | | 91 | | | | | |
| | | | | | | 103 | | | 106 | | | | | | |
| | | | | | | | 120 | 121 | | | | | | | |
| | | | | | | | 136 | 137 | | | | | | | |
| | | | | | | 151 | | | 154 | | | | | | |
| | | | | | 166 | | | | | 171 | | | | | |
| | | | | 181 | | | | | | | 188 | | | | |
| | | | 196 | | | | | | | | | 205 | | | |
| | | 211 | | | | | | | | | | | 222 | | |
| | 226 | | | | | | | | | | | | | 239 | |
| 241 | | | | | | | | | | | | | | | 256 |

Negative [A] Start with the bottom row, right to left in each row, going up. Begin with the number 5, count each square, only fill-in shaded squares. Helpful hints below.

| | | | | 252 | | | | | | | 245 | | | | |
|---|---|---|---|---|---|---|---|---|---|---|---|---|---|---|---|
| | | 238 | | | | | | | | | | | 227 | | |
| | 223 | | | | | | | | | | | | | 210 | |
| | | | | 203 | | | | | | 198 | | | | | |
| 192 | | | | | | | | | | | | | | | 177 |
| | | 173 | | | | | | | | | 164 | | | | |
| | | | | | | 153 | 152 | | | | | | | | |
| | | | | | 138 | | | 135 | | | | | | | |
| | | | | | 122 | | | 119 | | | | | | | |
| | | | | | | 105 | 104 | | | | | | | | |
| | | 93 | | | | | | | | | 84 | | | | |
| 80 | | | | | | | | | | | | | | | 65 |
| | | | | 59 | | | | | | 54 | | | | | |
| | 47 | | | | | | | | | | | | | 34 | |
| | | 30 | | | | | | | | | | | 19 | | |
| | | | | 12 | | | | | | 5 | | | | | |

Magic Square – Number Fill-ins [A] – Mathematical Formula 16 X 16 + 1 X 8 = 2056

# SHELSEAS MAGIC SQUARE II
## It Works

Positive [A] Start with the top row, left to right in each row, going down.  Begin with the number 1, count each square, only fill-in same shaded squares.  Diagonal are shown for the same sum of 2056.

Negative [A] Start with the bottom row, right to left in each row, going up.  Begin with the number 5, count each square, only fill-in same shaded squares.

Objective:  Combine both Positive & Negative number fill-ins [A].  Add each square of numbers in any row, column, or diagonal to proof the same sum of 2056 for a solved Magic Square.  Helpful hints below.

| | | | | | | | | | | | | | | | |
|---|---|---|---|---|---|---|---|---|---|---|---|---|---|---|---|
| 1 | | | | 252 | | | | | | | 245 | | | | 16 |
| | 18 | 238 | | | | | | | | | | | 227 | 31 | |
| | 223 | 35 | | | | | | | | | | | 46 | 210 | |
| | | | 52 | | 203 | | | | | 198 | | 61 | | | |
| 192 | | | | 69 | | | | | | | 76 | | | | 177 |
| | | | 173 | | 86 | | | | | 91 | | 164 | | | |
| | | | | | | 103 | 153 | 152 | 106 | | | | | | |
| | | | | | | 138 | 120 | 121 | 135 | | | | | | |
| | | | | | | 122 | 136 | 137 | 119 | | | | | | |
| | | | | | | 151 | 105 | 104 | 154 | | | | | | |
| | | | 93 | | 166 | | | | | 171 | | 84 | | | |
| 80 | | | | 181 | | | | | | | 188 | | | | 65 |
| | | | 196 | | 59 | | | | | 54 | | 205 | | | |
| | 47 | 211 | | | | | | | | | | | 222 | 34 | |
| | 226 | 30 | | | | | | | | | | | 19 | 239 | |
| 241 | | | | 12 | | | | | | | 5 | | | | 256 |

Magic Square
Number Fill-ins [A]
Mathematical Formula 16 X 16 + 1 X 8 = 2056

# SHELSEAS MAGIC SQUARE II
## It Works

Positive [A] Start with the top row, left to right in each row, going down.  Begin with the number 1, count each square, only fill-in shaded squares.  Diagonals are shown for the same sum of 4010.  Helpful hints below.

| 1 | | | | | | | | | | | | | | | | | | | 20 |
|---|---|---|---|---|---|---|---|---|---|---|---|---|---|---|---|---|---|---|---|
| | 22 | | | | | | | | | | | | | | | | | 39 | |
| | | 43 | | | | | | | | | | | | | | | 58 | | |
| | | | 64 | | | | | | | | | | | | | 77 | | | |
| | | | | 85 | | | | | | | | | | | 96 | | | | |
| | | | | | 106 | | | | | | | | | 115 | | | | | |
| | | | | | | 127 | | | | | | | 134 | | | | | | |
| | | | | | | | 148 | | | | | 153 | | | | | | | |
| | | | | | | | | 169 | | | 172 | | | | | | | | |
| | | | | | | | | | 190 | 191 | | | | | | | | | |
| | | | | | | | | | 210 | 211 | | | | | | | | | |
| | | | | | | | | 229 | | | 232 | | | | | | | | |
| | | | | | | | 248 | | | | | 253 | | | | | | | |
| | | | | | | 267 | | | | | | | 274 | | | | | | |
| | | | | | 286 | | | | | | | | | 295 | | | | | |
| | | | | 305 | | | | | | | | | | | 316 | | | | |
| | | | 324 | | | | | | | | | | | | | 337 | | | |
| | | 343 | | | | | | | | | | | | | | | 358 | | |
| | 362 | | | | | | | | | | | | | | | | | 379 | |
| 381 | | | | | | | | | | | | | | | | | | | 400 |

Magic Square
Number Fill-ins [A]
Mathematical Formula 20 X 20 + 1 X 10 = 4010

# SHELSEAS MAGIC SQUARE II
## It Works

Negative [A] Start with the bottom row, right to left in each row, going up.  Begin with the number 3, count each square, only fill-in shaded squares.  Helpful hints below.

| | | | | | | | | | | | | | | | | | | | |
|---|---|---|---|---|---|---|---|---|---|---|---|---|---|---|---|---|---|---|---|
| | | 398 | | | | | | | | | | | | | | | 383 | | |
| | | | | 375 | | | | | | | | | | 366 | | | | | |
| 360 | | | | | | | | | | | | | | | | | | | 341 |
| | | | | | | | | 331 | 330 | | | | | | | | | | |
| | | | | | 314 | | | | | | | | 307 | | | | | | |
| | 299 | | | | | | | | | | | | | | | | | 282 | |
| | | | 276 | | | | | | | | | | | | 265 | | | | |
| | | | | | | | 252 | | | 249 | | | | | | | | | |
| | | | | | | 233 | | | | | 228 | | | | | | | | |
| | | 217 | | | | | | | | | | | | | | 204 | | | |
| | | 197 | | | | | | | | | | | | | | 184 | | | |
| | | | | | | | 173 | | | | | 168 | | | | | | | |
| | | | | | | | | 152 | | 149 | | | | | | | | | |
| | | | | 136 | | | | | | | | | | | 125 | | | | |
| | 119 | | | | | | | | | | | | | | | | | 102 | |
| | | | | | | 94 | | | | | | | 87 | | | | | | |
| | | | | | | | | 71 | 70 | | | | | | | | | | |
| 60 | | | | | | | | | | | | | | | | | | | 41 |
| | | | | 35 | | | | | | | | | | 26 | | | | | |
| | | 18 | | | | | | | | | | | | | | | 3 | | |

## Magic Square
## Number Fill-ins [A]
## Mathematical Formula 20 X 20 + 1 X 10 = 4010

# SHELSEAS MAGIC SQUARE II
## It Works

Positive [A] Start with the top row, left to right in each row, going down. Begin with the number 1, count each square, only fill-in same shaded squares. Diagonal are shown for the same sum of 4010.

Negative [A] Start with the bottom row, right to left in each row, going up. Begin with the number 3, count each square, only fill-in same shaded squares.

Objective: Combine both Positive & Negative number fill-ins [A]. Add each square of numbers in any row, column, or diagonal to proof the same sum of 4010 for a solved Magic Square. Helpful hints below.

| 1 | 2 | 3 | 4 | 5 | 6 | 7 | 8 | 9 | 10 | 11 | 12 | 13 | 14 | 15 | 16 | 17 | 18 | 19 | 20 |
|---|---|---|---|---|---|---|---|---|----|----|----|----|----|----|----|----|----|----|----|
| 1 |  | 398 |  |  |  |  |  |  |  |  |  |  |  |  |  |  | 383 |  | 20 |
|  | 22 |  |  |  | 375 |  |  |  |  |  |  |  |  | 366 |  |  |  | 39 |  |
| 360 |  | 43 |  |  |  |  |  |  |  |  |  |  |  |  |  |  | 58 |  | 341 |
|  |  |  | 64 |  |  |  |  |  | 331 | 330 |  |  |  |  |  | 77 |  |  |  |
|  |  |  |  | 85 |  | 314 |  |  |  |  |  |  | 307 |  | 96 |  |  |  |  |
|  | 299 |  |  |  | 106 |  |  |  |  |  |  |  |  | 115 |  |  |  | 282 |  |
|  |  |  |  | 276 |  | 127 |  |  |  |  |  |  | 134 |  | 265 |  |  |  |  |
|  |  |  |  |  |  |  | 148 | 252 |  |  | 249 | 153 |  |  |  |  |  |  |  |
|  |  |  |  |  |  |  | 233 | 169 |  |  | 172 | 228 |  |  |  |  |  |  |  |
|  |  |  | 217 |  |  |  |  |  | 190 | 191 |  |  |  |  |  | 204 |  |  |  |
|  |  |  | 197 |  |  |  |  |  | 210 | 211 |  |  |  |  |  | 184 |  |  |  |
|  |  |  |  |  |  |  | 173 | 229 |  |  | 232 | 168 |  |  |  |  |  |  |  |
|  |  |  |  |  |  |  | 248 | 152 |  |  | 149 | 253 |  |  |  |  |  |  |  |
|  |  |  |  | 136 |  | 267 |  |  |  |  |  |  | 274 |  | 125 |  |  |  |  |
|  | 119 |  |  |  | 286 |  |  |  |  |  |  |  |  | 295 |  |  |  | 102 |  |
|  |  |  |  | 305 |  | 94 |  |  |  |  |  |  | 87 |  | 316 |  |  |  |  |
|  |  |  | 324 |  |  |  |  |  | 71 | 70 |  |  |  |  |  | 337 |  |  |  |
| 60 |  | 343 |  |  |  |  |  |  |  |  |  |  |  |  |  |  | 358 |  | 41 |
|  | 362 |  |  |  | 35 |  |  |  |  |  |  |  |  | 26 |  |  |  | 379 |  |
| 381 |  | 18 |  |  |  |  |  |  |  |  |  |  |  |  |  |  | 3 |  | 400 |

Magic Square – Number Fill-ins [A] – Mathematical Formula 20 X 20 + 1 X 10 = 4010

# SHELSEAS MAGIC SQUARE II
## It Work

Positive [A] Start with the top row, left to right in each row, going down. Begin with the number 1, count each square, only fill-in shaded squares. Helpful hints below.

| | | | | | | | | | | | | | | | | | | | | | | | |
|---|---|---|---|---|---|---|---|---|---|---|---|---|---|---|---|---|---|---|---|---|---|---|---|
| 1 | | | | | | | | | | | | | | | | | | | | | | | 24 |
| | | | 28 | | | | | | | | | | | | | | | | | 45 | | | |
| | | | | | | | 57 | | | | | | | | 64 | | | | | | | | |
| | | | | | | | | | 83 | | | 86 | | | | | | | | | | | |
| | 98 | | | | | | | | | | | | | | | | | | | | 119 | | |
| | | | | 125 | | | | | | | | | | | | | | 140 | | | | | |
| | | | | | | | 152 | | | | | | | | | 161 | | | | | | | |
| | | 171 | | | | | | | | | | | | | | | | | | | 190 | | |
| | | | | | | | | | 202 | | | | | 207 | | | | | | | | | |
| | | | | | 222 | | | | | | | | | | | | | 235 | | | | | |
| | | | | | | 247 | | | | | | | | | | | 258 | | | | | | |
| | | | | | | | | | | | 276 | 277 | | | | | | | | | | | |
| | | | | | | | | | | | 300 | 301 | | | | | | | | | | | |
| | | | | | | 319 | | | | | | | | | | | 330 | | | | | | |
| | | | | | 342 | | | | | | | | | | | | | 355 | | | | | |
| | | | | | | | | | 370 | | | | | 375 | | | | | | | | | |
| | | 387 | | | | | | | | | | | | | | | | | | | 406 | | |
| | | | | | | | 416 | | | | | | | | | 425 | | | | | | | |
| | | | | 437 | | | | | | | | | | | | | | 452 | | | | | |
| | 458 | | | | | | | | | | | | | | | | | | | | 479 | | |
| | | | | | | | | | | 491 | | | 494 | | | | | | | | | | |
| | | | | | | | | 513 | | | | | | | 520 | | | | | | | | |
| | | | 532 | | | | | | | | | | | | | | | | | 549 | | | |
| 553 | | | | | | | | | | | | | | | | | | | | | | | 576 |

Magic Square
Number Fill-ins [A]
Mathematical Formula 24 X 24 + 1 X 12 = 6924

# SHELSEAS MAGIC SQUARE II
## It Works

Negative [A] Start with the bottom row, right to left in each row, going up.  Begin with the number 4, count each square, only fill-in shaded squares.  Helpful hints below.

| 1 | 2 | 3 | 4 | 5 | 6 | 7 | 8 | 9 | 10 | 11 | 12 | 13 | 14 | 15 | 16 | 17 | 18 | 19 | 20 | 21 | 22 | 23 | 24 |
|---|---|---|---|---|---|---|---|---|----|----|----|----|----|----|----|----|----|----|----|----|----|----|----|
|  |  | 573 |  |  |  |  |  |  |  |  |  |  |  |  |  |  |  |  |  |  | 556 |  |  |
|  |  |  |  |  |  |  |  |  |  |  |  | 541 | 540 |  |  |  |  |  |  |  |  |  |  |
|  |  |  |  |  |  |  |  |  |  | 519 |  |  |  |  |  | 514 |  |  |  |  |  |  |  |
|  |  |  |  |  | 500 |  |  |  |  |  |  |  |  |  |  |  |  |  |  | 485 |  |  |  |
|  |  |  |  |  |  |  |  | 473 |  |  |  |  |  |  |  |  | 464 |  |  |  |  |  |  |
|  |  |  |  |  |  |  |  |  |  |  | 446 |  |  | 443 |  |  |  |  |  |  |  |  |  |
|  |  | 431 |  |  |  |  |  |  |  |  |  |  |  |  |  |  |  |  |  |  |  |  | 410 |
|  |  |  |  |  |  | 403 |  |  |  |  |  |  |  |  |  |  |  |  | 390 |  |  |  |  |
|  |  |  |  |  |  |  | 378 |  |  |  |  |  |  |  |  |  |  | 367 |  |  |  |  |  |
|  |  |  | 358 |  |  |  |  |  |  |  |  |  |  |  |  |  |  |  |  |  |  | 339 |  |
| 336 |  |  |  |  |  |  |  |  |  |  |  |  |  |  |  |  |  |  |  |  |  |  | 313 |
|  |  |  |  |  |  |  |  |  | 304 |  |  |  |  |  |  | 297 |  |  |  |  |  |  |  |
|  |  |  |  |  |  |  |  |  | 280 |  |  |  |  |  |  | 273 |  |  |  |  |  |  |  |
| 264 |  |  |  |  |  |  |  |  |  |  |  |  |  |  |  |  |  |  |  |  |  |  | 241 |
|  |  |  | 238 |  |  |  |  |  |  |  |  |  |  |  |  |  |  |  |  |  |  | 219 |  |
|  |  |  |  |  |  |  | 210 |  |  |  |  |  |  |  |  |  |  | 199 |  |  |  |  |  |
|  |  |  |  |  |  | 187 |  |  |  |  |  |  |  |  |  |  |  |  | 174 |  |  |  |  |
|  |  | 167 |  |  |  |  |  |  |  |  |  |  |  |  |  |  |  |  |  |  |  |  | 146 |
|  |  |  |  |  |  |  |  |  |  |  | 134 |  |  | 131 |  |  |  |  |  |  |  |  |  |
|  |  |  |  |  |  |  |  | 113 |  |  |  |  |  |  |  |  | 104 |  |  |  |  |  |  |
|  |  |  |  |  | 92 |  |  |  |  |  |  |  |  |  |  |  |  |  |  | 77 |  |  |  |
|  |  |  |  |  |  |  |  |  |  | 63 |  |  |  |  | 58 |  |  |  |  |  |  |  |  |
|  |  |  |  |  |  |  |  |  |  |  |  | 37 | 36 |  |  |  |  |  |  |  |  |  |  |
|  |  | 21 |  |  |  |  |  |  |  |  |  |  |  |  |  |  |  |  |  |  | 4 |  |  |

Magic Square
Number Fill-ins [A]
Mathematical Formula 24 X 24 + 1 X 12 = 6924

# SHELSEAS MAGIC SQUARE II
## It Works

Positive [A] Start with the top row, left to right in each row, going down.  Begin with the number 1, count each square, only fill-in same shaded squares.

Negative [A] Start with the bottom row, right to left in each row, going up.  Begin with the number 4, count each square, only fill-in same shaded squares.

Objective:  Combine both Positive & Negative number fill-ins [A].  Add each square of numbers in any row, column, or diagonal to proof the same sum of 6924 for a solved Magic Square.  Helpful hints below.

| 1 | 2 | 3 | 4 | 5 | 6 | 7 | 8 | 9 | 10 | 11 | 12 | 13 | 14 | 15 | 16 | 17 | 18 | 19 | 20 | 21 | 22 | 23 | 24 |
|---|---|---|---|---|---|---|---|---|---|---|---|---|---|---|---|---|---|---|---|---|---|---|---|
| 1 |  |  | 573 |  |  |  |  |  |  |  |  |  |  |  |  |  |  |  |  | 556 |  |  | 24 |
|  |  |  | 28 |  |  |  |  |  |  |  | 541 | 540 |  |  |  |  |  |  |  | 45 |  |  |  |
|  |  |  |  |  |  |  |  | 57 | 519 |  |  |  |  | 514 | 64 |  |  |  |  |  |  |  |  |
|  |  |  |  | 500 |  |  |  |  |  | 83 |  |  | 86 |  |  |  |  |  | 485 |  |  |  |  |
|  | 98 |  |  |  |  |  | 473 |  |  |  |  |  |  |  |  | 464 |  |  |  |  |  | 119 |  |
|  |  |  |  | 125 |  |  |  |  |  | 446 |  |  | 443 |  |  |  |  |  | 140 |  |  |  |  |
|  | 431 |  |  |  |  |  | 152 |  |  |  |  |  |  |  |  | 161 |  |  |  |  |  | 410 |  |
|  |  | 171 |  |  | 403 |  |  |  |  |  |  |  |  |  |  |  |  | 390 |  |  | 190 |  |  |
|  |  |  |  |  |  | 378 |  |  | 202 |  |  |  |  | 207 |  |  | 367 |  |  |  |  |  |  |
|  |  | 358 |  |  | 222 |  |  |  |  |  |  |  |  |  |  |  |  | 235 |  |  | 339 |  |  |
| 336 |  |  |  |  |  | 247 |  |  |  |  |  |  |  |  |  |  | 258 |  |  |  |  |  | 313 |
|  |  |  |  |  |  |  |  | 304 |  |  | 276 | 277 |  |  | 297 |  |  |  |  |  |  |  |  |
|  |  |  |  |  |  |  |  | 280 |  |  | 300 | 301 |  |  | 273 |  |  |  |  |  |  |  |  |
| 264 |  |  |  |  |  | 319 |  |  |  |  |  |  |  |  |  |  | 330 |  |  |  |  |  | 241 |
|  |  | 238 |  |  | 342 |  |  |  |  |  |  |  |  |  |  |  |  | 355 |  |  | 219 |  |  |
|  |  |  |  |  |  | 210 |  |  | 370 |  |  |  |  | 375 |  |  | 199 |  |  |  |  |  |  |
|  |  | 387 |  |  | 187 |  |  |  |  |  |  |  |  |  |  |  |  | 174 |  |  | 406 |  |  |
|  | 167 |  |  |  |  |  | 416 |  |  |  |  |  |  |  |  | 425 |  |  |  |  |  | 146 |  |
|  |  |  |  | 437 |  |  |  |  |  | 134 |  |  | 131 |  |  |  |  |  | 452 |  |  |  |  |
|  | 458 |  |  |  |  |  | 113 |  |  |  |  |  |  |  |  | 104 |  |  |  |  |  | 479 |  |
|  |  |  |  | 92 |  |  |  |  |  | 491 |  |  | 494 |  |  |  |  |  | 77 |  |  |  |  |
|  |  |  |  |  |  |  |  | 513 | 63 |  |  |  |  | 58 | 520 |  |  |  |  |  |  |  |  |
|  |  |  | 532 |  |  |  |  |  |  |  | 37 | 36 |  |  |  |  |  |  |  | 549 |  |  |  |
| 553 |  |  | 21 |  |  |  |  |  |  |  |  |  |  |  |  |  |  |  |  | 4 |  |  | 576 |

Magic Square – Number Fill-ins [A] – Mathematical Formula 24 X 24 + 1 X 12 = 6924

# Solved

# [A]

Mathematical Formulas

8 X 8 + 1 X 4 = 260

12 X 12 + 1 X 6 = 870

16 X 16 + 1 X 8 = 2056

20 X 20 + 1 X 10 = 4010

24 X 24 + 1 X 12 = 6924

Magic Square
Number Fill-ins [A]
Mathematical Formula

# SHELSEAS MAGIC SQUARE II
## It Works

### Positive [A]

| | 2 | 3 | | | 6 | 7 | |
|---|---|---|---|---|---|---|---|
| 9 | | | 12 | 13 | | | 16 |
| 17 | | 19 | | | 22 | | 24 |
| | 26 | | 28 | 29 | | 31 | |
| | 34 | | 36 | 37 | | 39 | |
| 41 | | 43 | | | 46 | | 48 |
| 49 | | | 52 | 53 | | | 56 |
| | 58 | 59 | | | 62 | 63 | |

### Negative [A]

| 64 | | | 61 | 60 | | | 57 |
|---|---|---|---|---|---|---|---|
| | 55 | 54 | | | 51 | 50 | |
| | 47 | | 45 | 44 | | 42 | |
| 40 | | 38 | | | 35 | | 33 |
| 32 | | 30 | | | 27 | | 25 |
| | 23 | | 21 | 20 | | 18 | |
| | 15 | 14 | | | 11 | 10 | |
| 8 | | | 5 | 4 | | | 1 |

Combination of both Positive & Negative number fill-ins [A] is shown for a solved Magic Square. Objective: Add each square of numbers in any row, column, or diagonal to proof the same sum of 260.

| 64 | 2 | 3 | 61 | 60 | 6 | 7 | 57 |
|---|---|---|---|---|---|---|---|
| 9 | 55 | 54 | 12 | 13 | 51 | 50 | 16 |
| 17 | 47 | 19 | 45 | 44 | 22 | 42 | 24 |
| 40 | 26 | 38 | 28 | 29 | 35 | 31 | 33 |
| 32 | 34 | 30 | 36 | 37 | 27 | 39 | 25 |
| 41 | 23 | 43 | 21 | 20 | 46 | 18 | 48 |
| 49 | 15 | 14 | 52 | 53 | 11 | 10 | 56 |
| 8 | 58 | 59 | 5 | 4 | 62 | 63 | 1 |

Magic Square
Number Fill-ins [A]
Mathematical Formula 8 X 8 + 1 X 4 = 260

# SHELSEAS MAGIC SQUARE II
## It Works

### Positive [A]

| 1 | 2 | 3 | | | | | | | 10 | 11 | 12 |
|---|---|---|---|---|---|---|---|---|---|---|---|
| 13 | 14 | | | | 18 | 19 | | | | 23 | 24 |
| 25 | | | 28 | 29 | | | 32 | 33 | | | 36 |
| | | 39 | 40 | 41 | | | 44 | 45 | 46 | | |
| | | 51 | 52 | | 54 | 55 | | 57 | 58 | | |
| | 62 | | | 65 | 66 | 67 | 68 | | | 71 | |
| | 74 | | | 77 | 78 | 79 | 80 | | | 83 | |
| | | 87 | 88 | | 90 | 91 | | 93 | 94 | | |
| | | 99 | 100 | 101 | | | 104 | 105 | 106 | | |
| 109 | | | 112 | 113 | | | 116 | 117 | | | 120 |
| 121 | 122 | | | | 126 | 127 | | | | 131 | 132 |
| 133 | 134 | 135 | | | | | | | 142 | 143 | 144 |

### Negative [A]

| | | | 141 | 140 | 139 | 138 | 137 | 136 | | | |
|---|---|---|---|---|---|---|---|---|---|---|---|
| | | 130 | 129 | 128 | | | 125 | 124 | 123 | | |
| | 119 | 118 | | | 115 | 114 | | | 111 | 110 | |
| 108 | 107 | | | | 103 | 102 | | | | 98 | 97 |
| 96 | 95 | | | 92 | | | 89 | | | 86 | 85 |
| 84 | | 82 | 81 | | | | | 76 | 75 | | 73 |
| 72 | | 70 | 69 | | | | | 64 | 63 | | 61 |
| 60 | 59 | | | 56 | | | 53 | | | 50 | 49 |
| 48 | 47 | | | | 43 | 42 | | | | 38 | 37 |
| | 35 | 34 | | | 31 | 30 | | | 27 | 26 | |
| | | 22 | 21 | 20 | | | 17 | 16 | 15 | | |
| | | | 9 | 8 | 7 | 6 | 5 | 4 | | | |

Combination of both Positive & Negative number fill-ins [A] is shown for a solved Magic Square. Objective: Add each square of numbers in any row, column, or diagonal to proof the same sum of 870.

| 1 | 2 | 3 | 141 | 140 | 139 | 138 | 137 | 136 | 10 | 11 | 12 |
|---|---|---|---|---|---|---|---|---|---|---|---|
| 13 | 14 | 130 | 129 | 128 | 18 | 19 | 125 | 124 | 123 | 23 | 24 |
| 25 | 119 | 118 | 28 | 29 | 115 | 114 | 32 | 33 | 111 | 110 | 36 |
| 108 | 107 | 39 | 40 | 41 | 103 | 102 | 44 | 45 | 46 | 98 | 97 |
| 96 | 95 | 51 | 52 | 92 | 54 | 55 | 89 | 57 | 58 | 86 | 85 |
| 84 | 62 | 82 | 81 | 65 | 66 | 67 | 68 | 76 | 75 | 71 | 73 |
| 72 | 74 | 70 | 69 | 77 | 78 | 79 | 80 | 64 | 63 | 83 | 61 |
| 60 | 59 | 87 | 88 | 56 | 90 | 91 | 53 | 93 | 94 | 50 | 49 |
| 48 | 47 | 99 | 100 | 101 | 43 | 42 | 104 | 105 | 106 | 38 | 37 |
| 109 | 35 | 34 | 112 | 113 | 31 | 30 | 116 | 117 | 27 | 26 | 120 |
| 121 | 122 | 22 | 21 | 20 | 126 | 127 | 17 | 16 | 15 | 131 | 132 |
| 133 | 134 | 135 | 9 | 8 | 7 | 6 | 5 | 4 | 142 | 143 | 144 |

Magic Square – Number Fill-ins [A] – Mathematical Formula 12 X 12 + 1 X 6 = 870

# SHELSEAS MAGIC SQUARE II
## It Works

Positive [A] Diagonals are shown for the same sum of 2056

| 1 | 2 | 3 | 4 | | | | | | | | | 13 | 14 | 15 | 16 |
|---|---|---|---|---|---|---|---|---|---|---|---|---|---|---|---|
| 17 | 18 | | | | | 23 | 24 | 25 | 26 | | | | | 31 | 32 |
| 33 | | 35 | 36 | 37 | | | | | | | 44 | 45 | 46 | | 48 |
| 49 | | 51 | 52 | | | 56 | 57 | | | | | 61 | 62 | | 64 |
| | | 67 | | 69 | 70 | 71 | | | 74 | 75 | 76 | | 78 | | |
| | | | | 85 | 86 | 87 | 88 | 89 | 90 | 91 | 92 | | | | |
| | 98 | | | 101 | 102 | 103 | | | 106 | 107 | 108 | | | 111 | |
| | 114 | | 116 | | 118 | | 120 | 121 | | 123 | | 125 | | 127 | |
| | 130 | | 132 | | 134 | | 136 | 137 | | 139 | | 141 | | 143 | |
| | 146 | | | 149 | 150 | 151 | | | 154 | 155 | 156 | | | 159 | |
| | | | | 165 | 166 | 167 | 168 | 169 | 170 | 171 | 172 | | | | |
| | | 179 | | 181 | 182 | 183 | | | 186 | 187 | 188 | | 190 | | |
| 193 | | 195 | 196 | | | | 200 | 201 | | | | 205 | 206 | | 208 |
| 209 | | 211 | 212 | 213 | | | | | | | 220 | 221 | 222 | | 224 |
| 225 | 226 | | | | | 231 | 232 | 233 | 234 | | | | | 239 | 240 |
| 241 | 242 | 243 | 244 | | | | | | | | | 253 | 254 | 255 | 256 |

## Negative [A]

| | | | | 252 | 251 | 250 | 249 | 248 | 247 | 246 | 245 | | | | |
|---|---|---|---|---|---|---|---|---|---|---|---|---|---|---|---|
| | | 238 | 237 | 236 | 235 | | | | | 230 | 229 | 228 | 227 | | |
| | 223 | | | | 219 | 218 | 217 | 216 | 215 | 214 | | | | 210 | |
| | 207 | | | 204 | 203 | 202 | | | 199 | 198 | 197 | | | 194 | |
| 192 | 191 | | 189 | | | | 185 | 184 | | | | 180 | | 178 | 177 |
| 176 | 175 | 174 | 173 | | | | | | | | | 164 | 163 | 162 | 161 |
| 160 | | 158 | 157 | | | | 153 | 152 | | | | 148 | 147 | | 145 |
| 144 | | 142 | | 140 | | 138 | | | 135 | | 133 | | 131 | | 129 |
| 128 | | 126 | | 124 | | 122 | | | 119 | | 117 | | 115 | | 113 |
| 112 | | 110 | 109 | | | | 105 | 104 | | | | 100 | 99 | | 97 |
| 96 | 95 | 94 | 93 | | | | | | | | | 84 | 83 | 82 | 81 |
| 80 | 79 | | 77 | | | | 73 | 72 | | | | 68 | | 66 | 65 |
| | 63 | | | 60 | 59 | 58 | | | 55 | 54 | 53 | | | 50 | |
| | 47 | | | | 43 | 42 | 41 | 40 | 39 | 38 | | | | 34 | |
| | | 30 | 29 | 28 | 27 | | | | | 22 | 21 | 20 | 19 | | |
| | | | | 12 | 11 | 10 | 9 | 8 | 7 | 6 | 5 | | | | |

## Magic Square
### Number Fill-ins [A]
### Mathematical Formula 16 X 16 + 1 X 8 = 2056

# SHELSEAS MAGIC SQUARE II
## It Works

Combination of both Positive & Negative number fill-ins [A] is shown for a solved Magic Square. Objective: Add each square of numbers in any row, column, or diagonal to proof the same sum of 2056.

| 1 | 2 | 3 | 4 | 252 | 251 | 250 | 249 | 248 | 247 | 246 | 245 | 13 | 14 | 15 | 16 |
|---|---|---|---|---|---|---|---|---|---|---|---|---|---|---|---|
| 17 | 18 | 238 | 237 | 236 | 235 | 23 | 24 | 25 | 26 | 230 | 229 | 228 | 227 | 31 | 32 |
| 33 | 223 | 35 | 36 | 37 | 219 | 218 | 217 | 216 | 215 | 214 | 44 | 45 | 46 | 210 | 48 |
| 49 | 207 | 51 | 52 | 204 | 203 | 202 | 56 | 57 | 199 | 198 | 197 | 61 | 62 | 194 | 64 |
| 192 | 191 | 67 | 189 | 69 | 70 | 71 | 185 | 184 | 74 | 75 | 76 | 180 | 78 | 178 | 177 |
| 176 | 175 | 174 | 173 | 85 | 86 | 87 | 88 | 89 | 90 | 91 | 92 | 164 | 163 | 162 | 161 |
| 160 | 98 | 158 | 157 | 101 | 102 | 103 | 153 | 152 | 106 | 107 | 108 | 148 | 147 | 111 | 145 |
| 144 | 114 | 142 | 116 | 140 | 118 | 138 | 120 | 121 | 135 | 123 | 133 | 125 | 131 | 127 | 129 |
| 128 | 130 | 126 | 132 | 124 | 134 | 122 | 136 | 137 | 119 | 139 | 117 | 141 | 115 | 143 | 113 |
| 112 | 146 | 110 | 109 | 149 | 150 | 151 | 105 | 104 | 154 | 155 | 156 | 100 | 99 | 159 | 97 |
| 96 | 95 | 94 | 93 | 165 | 166 | 167 | 168 | 169 | 170 | 171 | 172 | 84 | 83 | 82 | 81 |
| 80 | 79 | 179 | 77 | 181 | 182 | 183 | 73 | 72 | 186 | 187 | 188 | 68 | 190 | 66 | 65 |
| 193 | 63 | 195 | 196 | 60 | 59 | 58 | 200 | 201 | 55 | 54 | 53 | 205 | 206 | 50 | 208 |
| 209 | 47 | 211 | 212 | 213 | 43 | 42 | 41 | 40 | 39 | 38 | 220 | 221 | 222 | 34 | 224 |
| 225 | 226 | 30 | 29 | 28 | 27 | 231 | 232 | 233 | 234 | 22 | 21 | 20 | 19 | 239 | 240 |
| 241 | 242 | 243 | 244 | 12 | 11 | 10 | 9 | 8 | 7 | 6 | 5 | 253 | 254 | 255 | 256 |

# 2056

Magic Square
Number Fill-ins [A]
Mathematical Formula 16 X 16 + 1 X 8 = 2056

# SHELSEAS MAGIC SQUARE II
## It Works

### Positive [A] Diagonals are shown for the same sum of 4010

| 1 | 2 | | | 5 | | | | 9 | 10 | 11 | 12 | | | | 16 | | | 19 | 20 |
|---|---|---|---|---|---|---|---|---|---|---|---|---|---|---|---|---|---|---|---|
| 21 | 22 | | | 25 | | | | 29 | 30 | 31 | 32 | | | | 36 | | | 39 | 40 |
| | | 43 | 44 | | 46 | 47 | 48 | | | | | 53 | 54 | 55 | | 57 | 58 | | |
| | | 63 | 64 | | 66 | 67 | 68 | | | | | 73 | 74 | 75 | | 77 | 78 | | |
| 81 | 82 | | | 85 | | | | 89 | 90 | 91 | 92 | | | | 96 | | | 99 | 100 |
| | | 103 | 104 | | 106 | 107 | 108 | | | | | 113 | 114 | 115 | | 117 | 118 | | |
| | | 123 | 124 | | 126 | 127 | 128 | | | | | 133 | 134 | 135 | | 137 | 138 | | |
| | | 143 | 144 | | 146 | 147 | 148 | | | | | 153 | 154 | 155 | | 157 | 158 | | |
| 161 | 162 | | | 165 | | | | 169 | 170 | 171 | 172 | | | | 176 | | | 179 | 180 |
| 181 | 182 | | | 185 | | | | 189 | 190 | 191 | 192 | | | | 196 | | | 199 | 200 |
| 201 | 202 | | | 205 | | | | 209 | 210 | 211 | 212 | | | | 216 | | | 219 | 220 |
| 221 | 222 | | | 225 | | | | 229 | 230 | 231 | 232 | | | | 236 | | | 239 | 240 |
| | | 243 | 244 | | 246 | 247 | 248 | | | | | 253 | 254 | 255 | | 257 | 258 | | |
| | | 263 | 264 | | 266 | 267 | 268 | | | | | 273 | 274 | 275 | | 277 | 278 | | |
| | | 283 | 284 | | 286 | 287 | 288 | | | | | 293 | 294 | 295 | | 297 | 298 | | |
| 301 | 302 | | | 305 | | | | 309 | 310 | 311 | 312 | | | | 316 | | | 319 | 320 |
| | | 323 | 324 | | 326 | 327 | 328 | | | | | 333 | 334 | 335 | | 337 | 338 | | |
| | | 343 | 344 | | 346 | 347 | 348 | | | | | 353 | 354 | 355 | | 357 | 358 | | |
| 361 | 362 | | | 365 | | | | 369 | 370 | 371 | 372 | | | | 376 | | | 379 | 380 |
| 381 | 382 | | | 385 | | | | 389 | 390 | 391 | 392 | | | | 396 | | | 399 | 400 |

### Negative [A]

| | | 398 | 397 | | 395 | 394 | 393 | | | | | 388 | 387 | 386 | | 384 | 383 | | |
|---|---|---|---|---|---|---|---|---|---|---|---|---|---|---|---|---|---|---|---|
| | | 378 | 377 | | 375 | 374 | 373 | | | | | 368 | 367 | 366 | | 364 | 363 | | |
| 360 | 359 | | | 356 | | | | 352 | 351 | 350 | 349 | | | | 345 | | | 342 | 341 |
| 340 | 339 | | | 336 | | | | 332 | 331 | 330 | 329 | | | | 325 | | | 322 | 321 |
| | | 318 | 317 | | 315 | 314 | 313 | | | | | 308 | 307 | 306 | | 304 | 303 | | |
| 300 | 299 | | | 296 | | | | 292 | 291 | 290 | 289 | | | | 285 | | | 282 | 281 |
| 280 | 279 | | | 276 | | | | 272 | 271 | 270 | 269 | | | | 265 | | | 262 | 261 |
| 260 | 259 | | | 256 | | | | 252 | 251 | 250 | 249 | | | | 245 | | | 242 | 241 |
| | | 238 | 237 | | 235 | 234 | 233 | | | | | 228 | 227 | 226 | | 224 | 223 | | |
| | | 218 | 217 | | 215 | 214 | 213 | | | | | 208 | 207 | 206 | | 204 | 203 | | |
| | | 198 | 197 | | 195 | 194 | 193 | | | | | 188 | 187 | 186 | | 184 | 183 | | |
| | | 178 | 177 | | 175 | 174 | 173 | | | | | 168 | 167 | 166 | | 164 | 163 | | |
| 160 | 159 | | | 156 | | | | 152 | 151 | 150 | 149 | | | | 145 | | | 142 | 141 |
| 140 | 139 | | | 136 | | | | 132 | 131 | 130 | 129 | | | | 125 | | | 122 | 121 |
| 120 | 119 | | | 116 | | | | 112 | 111 | 110 | 109 | | | | 105 | | | 102 | 101 |
| | | 98 | 97 | | 95 | 94 | 93 | | | | | 88 | 87 | 86 | | 84 | 83 | | |
| 80 | 79 | | | 76 | | | | 72 | 71 | 70 | 69 | | | | 65 | | | 62 | 61 |
| 60 | 59 | | | 56 | | | | 52 | 51 | 50 | 49 | | | | 45 | | | 42 | 41 |
| | | 38 | 37 | | 35 | 34 | 33 | | | | | 28 | 27 | 26 | | 24 | 23 | | |
| | | 18 | 17 | | 15 | 14 | 13 | | | | | 8 | 7 | 6 | | 4 | 3 | | |

Magic Square – Number Fill-ins [A] – Mathematical Formula 20 X 20 + 1 X 10 = 4010

# SHELSEAS MAGIC SQUARE II
## It Works

Combination of both Positive & Negative number fill-ins [A] is shown for a solved Magic Square. Objective: Add each square of numbers in any row, column, or diagonal to proof the same sum of 4010.

| 1 | 2 | 398 | 397 | 5 | 395 | 394 | 393 | 9 | 10 | 11 | 12 | 388 | 387 | 386 | 16 | 384 | 383 | 19 | 20 |
|---|---|---|---|---|---|---|---|---|---|---|---|---|---|---|---|---|---|---|---|
| 21 | 22 | 378 | 377 | 25 | 375 | 374 | 373 | 29 | 30 | 31 | 32 | 368 | 367 | 366 | 36 | 364 | 363 | 39 | 40 |
| 360 | 359 | 43 | 44 | 356 | 46 | 47 | 48 | 352 | 351 | 350 | 349 | 53 | 54 | 55 | 345 | 57 | 58 | 342 | 341 |
| 340 | 339 | 63 | 64 | 336 | 66 | 67 | 68 | 332 | 331 | 330 | 329 | 73 | 74 | 75 | 325 | 77 | 78 | 322 | 321 |
| 81 | 82 | 318 | 317 | 85 | 315 | 314 | 313 | 89 | 90 | 91 | 92 | 308 | 307 | 306 | 96 | 304 | 303 | 99 | 100 |
| 300 | 299 | 103 | 104 | 296 | 106 | 107 | 108 | 292 | 291 | 290 | 289 | 113 | 114 | 115 | 285 | 117 | 118 | 282 | 281 |
| 280 | 279 | 123 | 124 | 276 | 126 | 127 | 128 | 272 | 271 | 270 | 269 | 133 | 134 | 135 | 265 | 137 | 138 | 262 | 261 |
| 260 | 259 | 143 | 144 | 256 | 146 | 147 | 148 | 252 | 251 | 250 | 249 | 153 | 154 | 155 | 245 | 157 | 158 | 242 | 241 |
| 161 | 162 | 238 | 237 | 165 | 235 | 234 | 233 | 169 | 170 | 171 | 172 | 228 | 227 | 226 | 176 | 224 | 223 | 179 | 180 |
| 181 | 182 | 218 | 217 | 185 | 215 | 214 | 213 | 189 | 190 | 191 | 192 | 208 | 207 | 206 | 196 | 204 | 203 | 199 | 200 |
| 201 | 202 | 198 | 197 | 205 | 195 | 194 | 193 | 209 | 210 | 211 | 212 | 188 | 187 | 186 | 216 | 184 | 183 | 219 | 220 |
| 221 | 222 | 178 | 177 | 225 | 175 | 174 | 173 | 229 | 230 | 231 | 232 | 168 | 167 | 166 | 236 | 164 | 163 | 239 | 240 |
| 160 | 159 | 243 | 244 | 156 | 246 | 247 | 248 | 152 | 151 | 150 | 149 | 253 | 254 | 255 | 145 | 257 | 258 | 142 | 141 |
| 140 | 139 | 263 | 264 | 136 | 266 | 267 | 268 | 132 | 131 | 130 | 129 | 273 | 274 | 275 | 125 | 277 | 278 | 122 | 121 |
| 120 | 119 | 283 | 284 | 116 | 286 | 287 | 288 | 112 | 111 | 110 | 109 | 293 | 294 | 295 | 105 | 297 | 298 | 102 | 101 |
| 301 | 302 | 98 | 97 | 305 | 95 | 94 | 93 | 309 | 310 | 311 | 312 | 88 | 87 | 86 | 316 | 84 | 83 | 319 | 320 |
| 80 | 79 | 323 | 324 | 76 | 326 | 327 | 328 | 72 | 71 | 70 | 69 | 333 | 334 | 335 | 65 | 337 | 338 | 62 | 61 |
| 60 | 59 | 343 | 344 | 56 | 346 | 347 | 348 | 52 | 51 | 50 | 49 | 353 | 354 | 355 | 45 | 357 | 358 | 42 | 41 |
| 361 | 362 | 38 | 37 | 365 | 35 | 34 | 33 | 369 | 370 | 371 | 372 | 28 | 27 | 26 | 376 | 24 | 23 | 379 | 380 |
| 381 | 382 | 18 | 17 | 385 | 15 | 14 | 13 | 389 | 390 | 391 | 392 | 8 | 7 | 6 | 396 | 4 | 3 | 399 | 400 |

# 4010

Magic Square
Number Fill-ins [A]
Mathematical Formula 20 X 20 + 1 X 10 = 4010

# SHELSEAS MAGIC SQUARE II
## It Works

### Positive [A]

| | | | | | | | | | | | | | | | | | | | | | | | |
|---|---|---|---|---|---|---|---|---|---|---|---|---|---|---|---|---|---|---|---|---|---|---|---|
| 1 | 2 | 3 | | | | | | | 10 | 11 | 12 | 13 | 14 | 15 | | | | | | | 22 | 23 | 24 |
| | | | 28 | 29 | 30 | 31 | 32 | 33 | | | | | | | 40 | 41 | 42 | 43 | 44 | 45 | | | |
| | | | 52 | 53 | 54 | 55 | 56 | 57 | | | | | | | 64 | 65 | 66 | 67 | 68 | 69 | | | |
| 73 | 74 | 75 | | | | | | | 82 | 83 | 84 | 85 | 86 | 87 | | | | | | | 94 | 95 | 96 |
| 97 | 98 | 99 | | | | | | | 106 | 107 | 108 | 109 | 110 | 111 | | | | | | | 118 | 119 | 120 |
| | | | 124 | 125 | 126 | 127 | 128 | 129 | | | | | | | 136 | 137 | 138 | 139 | 140 | 141 | | | |
| | | | 148 | 149 | 150 | 151 | 152 | 153 | | | | | | | 160 | 161 | 162 | 163 | 164 | 165 | | | |
| 169 | 170 | 171 | | | | | | | 178 | 179 | 180 | 181 | 182 | 183 | | | | | | | 190 | 191 | 192 |
| 193 | 194 | 195 | | | | | | | 202 | 203 | 204 | 205 | 206 | 207 | | | | | | | 214 | 215 | 216 |
| | | | 220 | 221 | 222 | 223 | 224 | 225 | | | | | | | 232 | 233 | 234 | 235 | 236 | 237 | | | |
| | | | 244 | 245 | 246 | 247 | 248 | 249 | | | | | | | 256 | 257 | 258 | 259 | 260 | 261 | | | |
| 265 | 266 | 267 | | | | | | | 274 | 275 | 276 | 277 | 278 | 279 | | | | | | | 286 | 287 | 288 |
| 289 | 290 | 291 | | | | | | | 298 | 299 | 300 | 301 | 302 | 303 | | | | | | | 310 | 311 | 312 |
| | | | 316 | 317 | 318 | 319 | 320 | 321 | | | | | | | 328 | 329 | 330 | 331 | 332 | 333 | | | |
| | | | 340 | 341 | 342 | 343 | 344 | 345 | | | | | | | 352 | 353 | 354 | 355 | 356 | 357 | | | |
| 361 | 362 | 363 | | | | | | | 370 | 371 | 372 | 373 | 374 | 375 | | | | | | | 382 | 383 | 384 |
| 385 | 386 | 387 | | | | | | | 394 | 395 | 396 | 397 | 398 | 399 | | | | | | | 406 | 407 | 408 |
| | | | 412 | 413 | 414 | 415 | 416 | 417 | | | | | | | 424 | 425 | 426 | 427 | 428 | 429 | | | |
| | | | 436 | 437 | 438 | 439 | 440 | 441 | | | | | | | 448 | 449 | 450 | 451 | 452 | 453 | | | |
| 457 | 458 | 459 | | | | | | | 466 | 467 | 468 | 469 | 470 | 471 | | | | | | | 478 | 479 | 480 |
| 481 | 482 | 483 | | | | | | | 490 | 491 | 492 | 493 | 494 | 495 | | | | | | | 502 | 503 | 504 |
| | | | 508 | 509 | 510 | 511 | 512 | 513 | | | | | | | 520 | 521 | 522 | 523 | 524 | 525 | | | |
| | | | 532 | 533 | 534 | 535 | 536 | 537 | | | | | | | 544 | 545 | 546 | 547 | 548 | 549 | | | |
| 553 | 554 | 555 | | | | | | | 562 | 563 | 564 | 565 | 566 | 567 | | | | | | | 574 | 575 | 576 |

### Negative [A]

| | | | | | | | | | | | | | | | | | | | | | | | |
|---|---|---|---|---|---|---|---|---|---|---|---|---|---|---|---|---|---|---|---|---|---|---|---|
| | | | 573 | 572 | 571 | 570 | 569 | 568 | | | | | | | 561 | 560 | 559 | 558 | 557 | 556 | | | |
| 552 | 551 | 550 | | | | | | | 543 | 542 | 541 | 540 | 539 | 538 | | | | | | | 531 | 530 | 529 |
| 528 | 527 | 526 | | | | | | | 519 | 518 | 517 | 516 | 515 | 514 | | | | | | | 507 | 506 | 505 |
| | | | 501 | 500 | 499 | 498 | 497 | 496 | | | | | | | 489 | 488 | 487 | 486 | 485 | 484 | | | |
| | | | 477 | 476 | 475 | 474 | 473 | 472 | | | | | | | 465 | 464 | 463 | 462 | 461 | 460 | | | |
| 456 | 455 | 454 | | | | | | | 447 | 446 | 445 | 444 | 443 | 442 | | | | | | | 435 | 434 | 433 |
| 432 | 431 | 430 | | | | | | | 423 | 422 | 421 | 420 | 419 | 418 | | | | | | | 411 | 410 | 409 |
| | | | 405 | 404 | 403 | 402 | 401 | 400 | | | | | | | 393 | 392 | 391 | 390 | 389 | 388 | | | |
| | | | 381 | 380 | 379 | 378 | 377 | 376 | | | | | | | 369 | 368 | 367 | 366 | 365 | 364 | | | |
| 360 | 359 | 358 | | | | | | | 351 | 350 | 349 | 348 | 347 | 346 | | | | | | | 339 | 338 | 337 |
| 336 | 335 | 334 | | | | | | | 327 | 326 | 325 | 324 | 323 | 322 | | | | | | | 315 | 314 | 313 |
| | | | 309 | 308 | 307 | 306 | 305 | 304 | | | | | | | 297 | 296 | 295 | 294 | 293 | 292 | | | |
| | | | 285 | 284 | 283 | 282 | 281 | 280 | | | | | | | 273 | 272 | 271 | 270 | 269 | 268 | | | |
| 264 | 263 | 262 | | | | | | | 255 | 254 | 253 | 252 | 251 | 250 | | | | | | | 243 | 242 | 241 |
| 240 | 239 | 238 | | | | | | | 231 | 230 | 229 | 228 | 227 | 226 | | | | | | | 219 | 218 | 217 |
| | | | 213 | 212 | 211 | 210 | 209 | 208 | | | | | | | 201 | 200 | 199 | 198 | 197 | 196 | | | |
| | | | 189 | 188 | 187 | 186 | 185 | 184 | | | | | | | 177 | 176 | 175 | 174 | 173 | 172 | | | |
| 168 | 167 | 166 | | | | | | | 159 | 158 | 157 | 156 | 155 | 154 | | | | | | | 147 | 146 | 145 |
| 144 | 143 | 142 | | | | | | | 135 | 134 | 133 | 132 | 131 | 130 | | | | | | | 123 | 122 | 121 |
| | | | 117 | 116 | 115 | 114 | 113 | 112 | | | | | | | 105 | 104 | 103 | 102 | 101 | 100 | | | |
| | | | 93 | 92 | 91 | 90 | 89 | 88 | | | | | | | 81 | 80 | 79 | 78 | 77 | 76 | | | |
| 72 | 71 | 70 | | | | | | | 63 | 62 | 61 | 60 | 59 | 58 | | | | | | | 51 | 50 | 49 |
| 48 | 47 | 46 | | | | | | | 39 | 38 | 37 | 36 | 35 | 34 | | | | | | | 27 | 26 | 25 |
| | | | 21 | 20 | 19 | 18 | 17 | 16 | | | | | | | 9 | 8 | 7 | 6 | 5 | 4 | | | |

Magic Square – Number Fill-ins [A] – Mathematical Formula 24 X 24 + 1 X 12 = 6924

# SHELSEAS MAGIC SQUARE II
## It Works

Combination of both Positive & Negative number fill-ins [A] is shown for a solved
Magic Square.  Objective:  Add each square of numbers in any row, column, or diagonal
to proof the same sum of 6924.

| 1 | 2 | 3 | 573 | 572 | 571 | 570 | 569 | 568 | 10 | 11 | 12 | 13 | 14 | 15 | 561 | 560 | 559 | 558 | 557 | 556 | 22 | 23 | 24 |
|---|---|---|-----|-----|-----|-----|-----|-----|----|----|----|----|----|----|-----|-----|-----|-----|-----|-----|----|----|----|
| 552 | 551 | 550 | 28 | 29 | 30 | 31 | 32 | 33 | 543 | 542 | 541 | 540 | 539 | 538 | 40 | 41 | 42 | 43 | 44 | 45 | 531 | 530 | 529 |
| 528 | 527 | 526 | 52 | 53 | 54 | 55 | 56 | 57 | 519 | 518 | 517 | 516 | 515 | 514 | 64 | 65 | 66 | 67 | 68 | 69 | 507 | 506 | 505 |
| 73 | 74 | 75 | 501 | 500 | 499 | 498 | 497 | 496 | 82 | 83 | 84 | 85 | 86 | 87 | 489 | 488 | 487 | 486 | 485 | 484 | 94 | 95 | 96 |
| 97 | 98 | 99 | 477 | 476 | 475 | 474 | 473 | 472 | 106 | 107 | 108 | 109 | 110 | 111 | 465 | 464 | 463 | 462 | 461 | 460 | 118 | 119 | 120 |
| 456 | 455 | 454 | 124 | 125 | 126 | 127 | 128 | 129 | 447 | 446 | 445 | 444 | 443 | 442 | 136 | 137 | 138 | 139 | 140 | 141 | 435 | 434 | 433 |
| 432 | 431 | 430 | 148 | 149 | 150 | 151 | 152 | 153 | 423 | 422 | 421 | 420 | 419 | 418 | 160 | 161 | 162 | 163 | 164 | 165 | 411 | 410 | 409 |
| 169 | 170 | 171 | 405 | 404 | 403 | 402 | 401 | 400 | 178 | 179 | 180 | 181 | 182 | 183 | 393 | 392 | 391 | 390 | 389 | 388 | 190 | 191 | 192 |
| 193 | 194 | 195 | 381 | 380 | 379 | 378 | 377 | 376 | 202 | 203 | 204 | 205 | 206 | 207 | 369 | 368 | 367 | 366 | 365 | 364 | 214 | 215 | 216 |
| 360 | 359 | 358 | 220 | 221 | 222 | 223 | 224 | 225 | 351 | 350 | 349 | 348 | 347 | 346 | 232 | 233 | 234 | 235 | 236 | 237 | 339 | 338 | 337 |
| 336 | 335 | 334 | 244 | 245 | 246 | 247 | 248 | 249 | 327 | 326 | 325 | 324 | 323 | 322 | 256 | 257 | 258 | 259 | 260 | 261 | 315 | 314 | 313 |
| 265 | 266 | 267 | 309 | 308 | 307 | 306 | 305 | 304 | 274 | 275 | 276 | 277 | 278 | 279 | 297 | 296 | 295 | 294 | 293 | 292 | 286 | 287 | 288 |
| 289 | 290 | 291 | 285 | 284 | 283 | 282 | 281 | 280 | 298 | 299 | 300 | 301 | 302 | 303 | 273 | 272 | 271 | 270 | 269 | 268 | 310 | 311 | 312 |
| 264 | 263 | 262 | 316 | 317 | 318 | 319 | 320 | 321 | 255 | 254 | 253 | 252 | 251 | 250 | 328 | 329 | 330 | 331 | 332 | 333 | 243 | 242 | 241 |
| 240 | 239 | 238 | 340 | 341 | 342 | 343 | 344 | 345 | 231 | 230 | 229 | 228 | 227 | 226 | 352 | 353 | 354 | 355 | 356 | 357 | 219 | 218 | 217 |
| 361 | 362 | 363 | 213 | 212 | 211 | 210 | 209 | 208 | 370 | 371 | 372 | 373 | 374 | 375 | 201 | 200 | 199 | 198 | 197 | 196 | 382 | 383 | 384 |
| 385 | 386 | 387 | 189 | 188 | 187 | 186 | 185 | 184 | 394 | 395 | 396 | 397 | 398 | 399 | 177 | 176 | 175 | 174 | 173 | 172 | 406 | 407 | 408 |
| 168 | 167 | 166 | 412 | 413 | 414 | 415 | 416 | 417 | 159 | 158 | 157 | 156 | 155 | 154 | 424 | 425 | 426 | 427 | 428 | 429 | 147 | 146 | 145 |
| 144 | 143 | 142 | 436 | 437 | 438 | 439 | 440 | 441 | 135 | 134 | 133 | 132 | 131 | 130 | 448 | 449 | 450 | 451 | 452 | 453 | 123 | 122 | 121 |
| 457 | 458 | 459 | 117 | 116 | 115 | 114 | 113 | 112 | 466 | 467 | 468 | 469 | 470 | 471 | 105 | 104 | 103 | 102 | 101 | 100 | 478 | 479 | 480 |
| 481 | 482 | 483 | 93 | 92 | 91 | 90 | 89 | 88 | 490 | 491 | 492 | 493 | 494 | 495 | 81 | 80 | 79 | 78 | 77 | 76 | 502 | 503 | 504 |
| 72 | 71 | 70 | 508 | 509 | 510 | 511 | 512 | 513 | 63 | 62 | 61 | 60 | 59 | 58 | 520 | 521 | 522 | 523 | 524 | 525 | 51 | 50 | 49 |
| 48 | 47 | 46 | 532 | 533 | 534 | 535 | 536 | 537 | 39 | 38 | 37 | 36 | 35 | 34 | 544 | 545 | 546 | 547 | 548 | 549 | 27 | 26 | 25 |
| 553 | 554 | 555 | 21 | 20 | 19 | 18 | 17 | 16 | 562 | 563 | 564 | 565 | 566 | 567 | 9 | 8 | 7 | 6 | 5 | 4 | 574 | 575 | 576 |

# 6924

Magic Square
Number Fill-ins [A]
Mathematical Formula 24 X 24 + 1 X 12 = 6924

The Number Fill-ins [B] Will Be Your Guideline
For A Solved Double Even – Magic Square.

# Number
# Fill-ins

# [B]

Positive [B] Start with the top row, right to left in each row, going down. Begin with the number shown hint, count each square, only fill-in same shaded squares.

Negative [B] Start with the bottom row, left to right in each row, going up. Begin with the number shown hint, count each square, only fill-in same shaded squares.

Objective: Combine both Positive & Negative number fill-ins [B] in each Magic Square. Add each square of numbers in any row, column, or diagonal to proof the same sum of integer for a solved Magic Square.

## Mathematical Formulas

$$8 \times 8 + 1 \times 4 = 260$$

$$12 \times 12 + 1 \times 6 = 870$$

$$16 \times 16 + 1 \times 8 = 2056$$

$$20 \times 20 + 1 \times 10 = 4010$$

$$24 \times 24 + 1 \times 12 = 6924$$

# SHELSEAS MAGIC SQUARE II
## It Works

Positive [B] Start with the top row, right to left in each row, going down. Begin with the number 1, count each square, only fill-in shaded squares.

| 4 |    |    | 1  |
|---|----|----|----|
|   | 7  | 6  |    |
|   | 11 | 10 |    |
| 16|    |    | 13 |

Negative [B] Start with the bottom row, left to right in each row, going up. Begin with the number 2, count each square, only fill-in shaded squares.

|   | 14 | 15 |    |
|---|----|----|----|
| 9 |    |    | 12 |
| 5 |    |    | 8  |
|   | 2  | 3  |    |

Combination of both Positive & Negative number fill-ins [B] is shown for a solved Magic Square. Objective: Add each square of numbers in any row, column, or diagonal to proof the same sum of 34.

| 4  | 14 | 15 | 1  |
|----|----|----|----|
| 9  | 7  | 6  | 12 |
| 5  | 11 | 10 | 8  |
| 16 | 2  | 3  | 13 |

Magic Square
Number Fill-ins [B]
Mathematical Formula 4 X 4 + 1 X 2 = 34

# SHELSEAS MAGIC SQUARE II
## It Works

Positive [B] Start with the top row, right to left in each row, going down. Begin with the number 1, count each square, only fill-in shaded squares. Diagonals are shown for the same sum of 260. Helpful hints below.

| 8 |  |  |  |  |  |  | 1 |
|---|---|---|---|---|---|---|---|
|  | 15 |  |  |  |  | 10 |  |
|  |  | 22 |  |  | 19 |  |  |
|  |  |  | 29 | 28 |  |  |  |
|  |  |  | 37 | 36 |  |  |  |
|  |  | 46 |  |  | 43 |  |  |
|  | 55 |  |  |  |  | 50 |  |
| 64 |  |  |  |  |  |  | 57 |

Negative [B] Start with the bottom row, left to right in each row, going up. Begin with the number 2, count each square, only fill-in shaded squares. Helpful hints below.

|  | 58 |  |  |  |  | 63 |  |
|---|---|---|---|---|---|---|---|
| 49 |  |  |  |  |  |  | 56 |
|  |  | 44 | 45 |  |  |  |  |
|  |  | 35 |  |  | 38 |  |  |
|  |  | 27 |  |  | 30 |  |  |
|  |  |  | 20 | 21 |  |  |  |
| 9 |  |  |  |  |  |  | 16 |
|  | 2 |  |  |  |  | 7 |  |

Objective: Combine both Positive & Negative number fill-ins [B]. Add each square of numbers in any row, column, or diagonal to proof the same sum of 260 for a solved Magic Square. Helpful hints below.

| 8 | 58 |  |  |  |  | 63 | 1 |
|---|---|---|---|---|---|---|---|
| 49 | 15 |  |  |  |  | 10 | 56 |
|  |  | 22 | 44 | 45 | 19 |  |  |
|  |  | 35 | 29 | 28 | 38 |  |  |
|  |  | 27 | 37 | 36 | 30 |  |  |
|  |  | 46 | 20 | 21 | 43 |  |  |
| 9 | 55 |  |  |  |  | 50 | 16 |
| 64 | 2 |  |  |  |  | 7 | 57 |

Magic Square – Number Fill-ins [B] – Mathematical Formula 8 X 8 + 1 X 4 = 260

# SHELSEAS MAGIC SQUARE II
## It Works

Positive [B] Start with the top row, right to left in each row, going down. Begin with the number 1, count each square, only fill-in shaded squares. Diagonals are shown for the same sum of 870. Helpful hints below

| | | | | | | | | | | | |
|---|---|---|---|---|---|---|---|---|---|---|---|
| 12 | | | | | | | | | | | 1 |
| | 23 | | | | | | | | | 14 | |
| | | 34 | | | | | | | 27 | | |
| | | | 45 | | | | | 40 | | | |
| | | | | 56 | | | 53 | | | | |
| | | | | | 67 | 66 | | | | | |
| | | | | | 79 | 78 | | | | | |
| | | | | 92 | | | 89 | | | | |
| | | | 105 | | | | | 100 | | | |
| | | 118 | | | | | | | 111 | | |
| | 131 | | | | | | | | | 122 | |
| 144 | | | | | | | | | | | 133 |

Negative [B] Start with the bottom row, left to right in each row, going up. Begin with the number 3, count each square, only fill-in shaded squares. Helpful hints below.

| | | | | | | | | | | | |
|---|---|---|---|---|---|---|---|---|---|---|---|
| | | 135 | | | | | | | 142 | | |
| | | | 124 | | | | | 129 | | | |
| | | | | 113 | | | 116 | | | | |
| | | | | | 102 | 103 | | | | | |
| 85 | | | | | | | | | | | 96 |
| | 74 | | | | | | | | | 83 | |
| | 62 | | | | | | | | | 71 | |
| 49 | | | | | | | | | | | 60 |
| | | | | | 42 | 43 | | | | | |
| | | | | 29 | | | 32 | | | | |
| | | | 16 | | | | | 21 | | | |
| | | 3 | | | | | | | 10 | | |

Magic Square
Number Fill-ins [B]
Mathematical Formula 12 X 12 + 1 X 6 = 870

# SHELSEAS MAGIC SQUARE II
## It Works

Positive [B] Start with the top row, right to left in each row, going down. Begin with the number 1, count each square, only fill-in same shaded squares. Diagonals are shown for the same sum of 870.

Negative [B] Start with the bottom row, left to right in each row, going up. Begin with the number 3, count each square, only fill-in same shaded squares.

Objective: Combine both Positive & Negative number fill-ins [B]. Add each square of numbers in any row, column, or diagonal to proof the same sum of 870 for a solved Magic Square. Helpful hints below.

| | | | | | | | | | | | |
|---|---|---|---|---|---|---|---|---|---|---|---|
| 12 | | 135 | | | | | | | 142 | | 1 |
| | 23 | | 124 | | | | | 129 | | 14 | |
| | | 34 | | 113 | | | 116 | | 27 | | |
| | | | 45 | | 102 | 103 | | 40 | | | |
| 85 | | | | 56 | | | 53 | | | | 96 |
| | 74 | | | | 67 | 66 | | | | 83 | |
| | 62 | | | | 79 | 78 | | | | 71 | |
| 49 | | | | 92 | | | 89 | | | | 60 |
| | | | 105 | | 42 | 43 | | 100 | | | |
| | | 118 | | 29 | | | 32 | | 111 | | |
| | 131 | | 16 | | | | | 21 | | 122 | |
| 144 | | 3 | | | | | | | 10 | | 133 |

Magic Square
Number Fill-ins [B]
Mathematical Formula 12 X 12 + 1 X 6 = 870

# SHELSEAS MAGIC SQUARE II
## It Works

Positive [B] Start with the top row, right to left in each row, going down. Begin with the number 1, count each square, only fill-in shaded squares. Helpful hints below.

| | | | | | | | | | | | | | | | |
|---|---|---|---|---|---|---|---|---|---|---|---|---|---|---|---|
| 16 | | | | | | | | | | | | | | | 1 |
| | 31 | | | | | | | | | | | | | 18 | |
| | | | | | | 42 | | | 39 | | | | | | |
| | | | | 60 | | | | | | | 53 | | | | |
| | | | 77 | | | | | | | | | 68 | | | |
| | | | | | 91 | | | | | 86 | | | | | |
| | | 110 | | | | | | | | | | | 99 | | |
| | | | | | | 121 | 120 | | | | | | | | |
| | | | | | | 137 | 136 | | | | | | | | |
| | | 158 | | | | | | | | | | | 147 | | |
| | | | | | 171 | | | | | 166 | | | | | |
| | | | 189 | | | | | | | | | 180 | | | |
| | | | | 204 | | | | | | | 197 | | | | |
| | | | | | | 218 | | | 215 | | | | | | |
| | 239 | | | | | | | | | | | | | 226 | |
| 256 | | | | | | | | | | | | | | | 241 |

Negative [B] Start with the bottom row, left to right in each row, going up. Begin with the number 4, count each square, only fill-in shaded squares. Helpful hints below.

| | | | | | | | | | | | | | | | |
|---|---|---|---|---|---|---|---|---|---|---|---|---|---|---|---|
| | | | 244 | | | | | | | | | 253 | | | |
| | | | | | | 231 | | | 234 | | | | | | |
| | | | | | 214 | | | | | 219 | | | | | |
| 193 | | | | | | | | | | | | | | | 208 |
| | | | | | | | 184 | 185 | | | | | | | |
| | | | 163 | | | | | | | | | | 174 | | |
| | 146 | | | | | | | | | | | | | 159 | |
| | | | | 133 | | | | | | 140 | | | | | |
| | | | | 117 | | | | | | 124 | | | | | |
| | | 98 | | | | | | | | | | | 111 | | |
| | | 83 | | | | | | | | | | | 94 | | |
| | | | | | | | 72 | 73 | | | | | | | |
| 49 | | | | | | | | | | | | | | | 64 |
| | | | | | 38 | | | | | 43 | | | | | |
| | | | | | | 23 | | | 26 | | | | | | |
| | | | 4 | | | | | | | | | 13 | | | |

Magic Square – Number Fill-ins [B] – Mathematical Formula 16 X 16 + 1 X 8 = 2056

# SHELSEAS MAGIC SQUARE II
## It Works

Positive [B] Start with the top row, right to left in each row, going down. Begin with the number 1, count each square, only fill-in same shaded squares.

Negative [B] Start with the bottom row, left to right in each row, going up. Begin with the number 4, count each square, only fill-in same shaded squares.

Objective: Combine both Positive & Negative number fill-ins [B]. Add each square of numbers in any row, column, or diagonal to proof the same sum of 2056 for a solved Magic Square. Helpful hints below.

| | | | | | | | | | | | | | | | |
|---|---|---|---|---|---|---|---|---|---|---|---|---|---|---|---|
| 16 | | | 244 | | | | | | | | | 253 | | | 1 |
| | 31 | | | | | 231 | | | 234 | | | | | 18 | |
| | | | | | 214 | 42 | | | 39 | 219 | | | | | |
| 193 | | | | 60 | | | | | | | 53 | | | | 208 |
| | | | 77 | | | | 184 | 185 | | | | 68 | | | |
| | | 163 | | | 91 | | | | | 86 | | | 174 | | |
| | 146 | 110 | | | | | | | | | | | 99 | 159 | |
| | | | | 133 | | | 121 | 120 | | | 140 | | | | |
| | | | | 117 | | | 137 | 136 | | | 124 | | | | |
| | 98 | 158 | | | | | | | | | | | 147 | 111 | |
| | | 83 | | | 171 | | | | | 166 | | | 94 | | |
| | | | 189 | | | | 72 | 73 | | | | 180 | | | |
| 49 | | | | 204 | | | | | | | 197 | | | | 64 |
| | | | | | 38 | 218 | | | 215 | 43 | | | | | |
| | 239 | | | | | 23 | | | 26 | | | | | 226 | |
| 256 | | | 4 | | | | | | | | | 13 | | | 241 |

Magic Square
Number Fill-ins [B]
Mathematical Formula16 X 16 + 1 X 8 = 2056

# SHELSEAS MAGIC SQUARE II
## It Works

Positive [B] Start with the top row, right to left in each row, going down. Begin with the number 1, count each square, only fill-in shaded squares. Helpful hints below.

| 1 | 2 | 3 | 4 | 5 | 6 | 7 | 8 | 9 | 10 | 11 | 12 | 13 | 14 | 15 | 16 | 17 | 18 | 19 | 20 |
|---|---|---|---|---|---|---|---|---|----|----|----|----|----|----|----|----|----|----|----|
| 20 | | | | | | | | | | | | | | | | | | | 1 |
| | | | | | | | | 32 | | | 29 | | | | | | | | |
| | | 58 | | | | | | | | | | | | | | | 43 | | |
| | | | | 76 | | | | | | | | | | | 65 | | | | |
| | | | 97 | | | | | | | | | | | | | 84 | | | |
| | | | | | | 114 | | | | | | | 107 | | | | | | |
| | | | | | 135 | | | | | | | | | 126 | | | | | |
| | | | | | | | 153 | | | | | 148 | | | | | | | |
| | 179 | | | | | | | | | | | | | | | | | 162 | |
| | | | | | | | | | 191 | 190 | | | | | | | | | |
| | | | | | | | | | 211 | 210 | | | | | | | | | |
| | 239 | | | | | | | | | | | | | | | | | 222 | |
| | | | | | | | 253 | | | | | 248 | | | | | | | |
| | | | | | 275 | | | | | | | | | 266 | | | | | |
| | | | | | | 294 | | | | | | | 287 | | | | | | |
| | | | 317 | | | | | | | | | | | | | 304 | | | |
| | | | | 336 | | | | | | | | | | | 325 | | | | |
| | | 358 | | | | | | | | | | | | | | | 343 | | |
| | | | | | | | | 372 | | | 369 | | | | | | | | |
| 400 | | | | | | | | | | | | | | | | | | | 381 |

Magic Square
Number Fill-ins [B]
Mathematical Formula 20 X 20 + 1 X 10 = 4010

# SHELSEAS MAGIC SQUARE II
## It Works

Negative [B] Start with the bottom row, left to right in each row, going up. Begin with the number 5, count each square, only fill-in shaded squares. Helpful hints below.

| | | | | | | | | | | | | | | | | | | | |
|---|---|---|---|---|---|---|---|---|---|---|---|---|---|---|---|---|---|---|---|
| | | | | 385 | | | | | | | | | | 396 | | | | | |
| | | | | | | | | 370 | 371 | | | | | | | | | | |
| | | | | | 346 | | | | | | | | 355 | | | | | | |
| | | | 324 | | | | | | | | | | | | 337 | | | | |
| 301 | | | | | | | | | | | | | | | | | | | 320 |
| | | 283 | | | | | | | | | | | | | | 298 | | | |
| | | | | | | | 268 | | | | 273 | | | | | | | | |
| | | | | | | 247 | | | | | | 254 | | | | | | | |
| | | | | | | | 229 | | | 232 | | | | | | | | | |
| | 202 | | | | | | | | | | | | | | | 219 | | | |
| | 182 | | | | | | | | | | | | | | | 199 | | | |
| | | | | | | | 169 | | | 172 | | | | | | | | | |
| | | | | | | 147 | | | | | | 154 | | | | | | | |
| | | | | | | | 128 | | | | 133 | | | | | | | | |
| | | 103 | | | | | | | | | | | | | 118 | | | | |
| 81 | | | | | | | | | | | | | | | | | | | 100 |
| | | | 64 | | | | | | | | | | | 77 | | | | | |
| | | | | | 46 | | | | | | | | 55 | | | | | | |
| | | | | | | | | 30 | 31 | | | | | | | | | | |
| | | | | 5 | | | | | | | | | | 16 | | | | | |

Magic Square
Number Fill-ins [B]
Mathematical Formula 20 X 20 + 1 X 10 = 4010

# SHELSEAS MAGIC SQUARE II
## It Works

Positive [B] Start with the top row, right to left in each row, going down.  Begin with the number 1, count each square, only fill-in same shaded squares.

Negative [B] Start with the bottom row, left to right in each row, going up.  Begin with the number 5, count each square, only fill-in same shaded squares.

Objective:  Combine both Positive & Negative number fill-ins [B].  Add each square of numbers in any row, column, or diagonal to proof the same sum of 4010 for a solved Magic Square.  Helpful hints below.

|  |  |  |  |  |  |  |  |  |  |  |  |  |  |  |  |  |  |  |  |
|---|---|---|---|---|---|---|---|---|---|---|---|---|---|---|---|---|---|---|---|
| 20 |  |  |  | 385 |  |  |  |  |  |  |  |  |  |  | 396 |  |  |  | 1 |
|  |  |  |  |  |  |  |  | 32 | 370 | 371 | 29 |  |  |  |  |  |  |  |  |
|  |  | 58 |  |  | 346 |  |  |  |  |  |  |  |  | 355 |  |  | 43 |  |  |
|  |  |  | 324 | 76 |  |  |  |  |  |  |  |  |  |  | 65 | 337 |  |  |  |
| 301 |  |  | 97 |  |  |  |  |  |  |  |  |  |  |  |  | 84 |  |  | 320 |
|  |  | 283 |  |  |  | 114 |  |  |  |  |  |  | 107 |  |  |  | 298 |  |  |
|  |  |  |  |  | 135 |  | 268 |  |  |  |  | 273 |  | 126 |  |  |  |  |  |
|  |  |  |  |  |  | 247 | 153 |  |  |  |  | 148 | 254 |  |  |  |  |  |  |
|  | 179 |  |  |  |  |  |  | 229 |  |  | 232 |  |  |  |  |  |  | 162 |  |
|  | 202 |  |  |  |  |  |  |  | 191 | 190 |  |  |  |  |  |  |  | 219 |  |
|  | 182 |  |  |  |  |  |  |  | 211 | 210 |  |  |  |  |  |  |  | 199 |  |
|  | 239 |  |  |  |  |  |  | 169 |  |  | 172 |  |  |  |  |  |  | 222 |  |
|  |  |  |  |  |  | 147 | 253 |  |  |  |  | 248 | 154 |  |  |  |  |  |  |
|  |  |  |  |  | 275 |  | 128 |  |  |  |  | 133 |  | 266 |  |  |  |  |  |
|  |  | 103 |  |  |  | 294 |  |  |  |  |  |  | 287 |  |  |  | 118 |  |  |
| 81 |  |  | 317 |  |  |  |  |  |  |  |  |  |  |  |  | 304 |  |  | 100 |
|  |  |  | 64 | 336 |  |  |  |  |  |  |  |  |  |  | 325 | 77 |  |  |  |
|  |  | 358 |  |  | 46 |  |  |  |  |  |  |  |  | 55 |  |  | 343 |  |  |
|  |  |  |  |  |  |  |  | 372 | 30 | 31 | 369 |  |  |  |  |  |  |  |  |
| 400 |  |  |  | 5 |  |  |  |  |  |  |  |  |  |  | 16 |  |  |  | 381 |

Magic Square – Number Fill-ins [B] – Mathematical Formula 20 X 20 + 1 X 10 = 4010

# SHELSEAS MAGIC SQUARE II
## It Works

Positive [B] Start with the top row, right to left in each row, going down.  Begin with the number 1, count each square, only fill-in shaded squares.  Helpful hints below.

| 1 | 2 | 3 | 4 | 5 | 6 | 7 | 8 | 9 | 10 | 11 | 12 | 13 | 14 | 15 | 16 | 17 | 18 | 19 | 20 | 21 | 22 | 23 | 24 |
|---|---|---|---|---|---|---|---|---|----|----|----|----|----|----|----|----|----|----|----|----|----|----|----|
| 24 |  |  |  |  |  |  |  |  |  |  |  |  |  |  |  |  |  |  |  |  |  |  | 1 |
|  | 47 |  |  |  |  |  |  |  |  |  |  |  |  |  |  |  |  |  |  |  |  | 26 |  |
|  |  | 70 |  |  |  |  |  |  |  |  |  |  |  |  |  |  |  |  |  |  | 51 |  |  |
|  |  |  | 93 |  |  |  |  |  |  |  |  |  |  |  |  |  |  |  |  | 76 |  |  |  |
|  |  |  |  | 116 |  |  |  |  |  |  |  |  |  |  |  |  |  |  | 101 |  |  |  |  |
|  |  |  |  |  |  | 138 |  |  |  |  |  |  |  |  |  |  | 127 |  |  |  |  |  |  |
|  |  |  |  |  | 163 |  |  |  |  |  |  |  |  |  |  |  |  | 150 |  |  |  |  |  |
|  |  |  |  |  |  |  |  |  |  | 182 |  |  | 179 |  |  |  |  |  |  |  |  |  |  |
|  |  |  |  |  |  |  |  |  | 207 |  |  |  |  | 202 |  |  |  |  |  |  |  |  |  |
|  |  |  |  |  |  |  |  | 232 |  |  |  |  |  |  | 225 |  |  |  |  |  |  |  |  |
|  |  |  |  |  |  |  | 257 |  |  |  |  |  |  |  |  | 248 |  |  |  |  |  |  |  |
|  |  |  |  |  |  |  |  |  |  |  | 277 | 276 |  |  |  |  |  |  |  |  |  |  |  |
|  |  |  |  |  |  |  |  |  |  |  | 301 | 300 |  |  |  |  |  |  |  |  |  |  |  |
|  |  |  |  |  |  |  | 329 |  |  |  |  |  |  |  |  | 320 |  |  |  |  |  |  |  |
|  |  |  |  |  |  |  |  | 352 |  |  |  |  |  |  | 345 |  |  |  |  |  |  |  |  |
|  |  |  |  |  |  |  |  |  | 375 |  |  |  |  | 370 |  |  |  |  |  |  |  |  |  |
|  |  |  |  |  |  |  |  |  |  | 398 |  |  | 395 |  |  |  |  |  |  |  |  |  |  |
|  |  |  |  |  | 427 |  |  |  |  |  |  |  |  |  |  |  |  | 414 |  |  |  |  |  |
|  |  |  |  |  |  | 450 |  |  |  |  |  |  |  |  |  |  | 439 |  |  |  |  |  |  |
|  |  |  |  | 476 |  |  |  |  |  |  |  |  |  |  |  |  |  |  | 461 |  |  |  |  |
|  |  |  | 501 |  |  |  |  |  |  |  |  |  |  |  |  |  |  |  |  | 484 |  |  |  |
|  |  | 526 |  |  |  |  |  |  |  |  |  |  |  |  |  |  |  |  |  |  | 507 |  |  |
|  | 551 |  |  |  |  |  |  |  |  |  |  |  |  |  |  |  |  |  |  |  |  | 530 |  |
| 576 |  |  |  |  |  |  |  |  |  |  |  |  |  |  |  |  |  |  |  |  |  |  | 553 |

Magic Square
Number Fill-ins [B]
Mathematical Formula 24 X 24 + 1 X 12 = 6924

# SHELSEAS MAGIC SQUARE II
## It Works

Negative [B] Start with the bottom row, left to right in each row, going up. Begin with the number 6, count each square, only fill-in shaded squares. Helpful hints below.

| 1 | 2 | 3 | 4 | 5 | 6 | 7 | 8 | 9 | 10 | 11 | 12 | 13 | 14 | 15 | 16 | 17 | 18 | 19 | 20 | 21 | 22 | 23 | 24 |
|---|---|---|---|---|---|---|---|---|----|----|----|----|----|----|----|----|----|----|----|----|----|----|----|
|  |  |  |  |  | 558 |  |  |  |  |  |  |  |  |  |  |  |  |  | 571 |  |  |  |  |
|  |  |  | 532 |  |  |  |  |  |  |  |  |  |  |  |  |  |  |  |  |  | 549 |  |  |
|  |  |  |  |  |  |  |  |  |  |  | 516 | 517 |  |  |  |  |  |  |  |  |  |  |  |
|  | 482 |  |  |  |  |  |  |  |  |  |  |  |  |  |  |  |  |  |  |  |  |  | 503 |
|  |  |  |  |  |  |  |  |  | 466 |  |  |  |  | 471 |  |  |  |  |  |  |  |  |  |
| 433 |  |  |  |  |  |  |  |  |  |  |  |  |  |  |  |  |  |  |  |  |  |  | 456 |
|  |  |  |  |  |  | 415 |  |  |  |  |  |  |  |  |  |  |  | 426 |  |  |  |  |  |
|  |  |  |  |  |  |  |  | 393 |  |  |  |  |  |  |  | 400 |  |  |  |  |  |  |  |
|  |  |  |  |  |  |  | 368 |  |  |  |  |  |  |  |  |  | 377 |  |  |  |  |  |  |
|  |  |  |  | 341 |  |  |  |  |  |  |  |  |  |  |  |  |  |  |  | 356 |  |  |  |
|  |  |  |  |  |  |  |  |  |  | 323 |  |  | 326 |  |  |  |  |  |  |  |  |  |  |
|  |  | 291 |  |  |  |  |  |  |  |  |  |  |  |  |  |  |  |  |  |  |  | 310 |  |
|  |  | 267 |  |  |  |  |  |  |  |  |  |  |  |  |  |  |  |  |  |  |  | 286 |  |
|  |  |  |  |  |  |  |  |  |  | 251 |  |  | 254 |  |  |  |  |  |  |  |  |  |  |
|  |  |  |  | 221 |  |  |  |  |  |  |  |  |  |  |  |  |  |  |  | 236 |  |  |  |
|  |  |  |  |  |  |  | 200 |  |  |  |  |  |  |  |  |  | 209 |  |  |  |  |  |  |
|  |  |  |  |  |  |  |  | 177 |  |  |  |  |  |  |  | 184 |  |  |  |  |  |  |  |
|  |  |  |  |  |  | 151 |  |  |  |  |  |  |  |  |  |  |  | 162 |  |  |  |  |  |
| 121 |  |  |  |  |  |  |  |  |  |  |  |  |  |  |  |  |  |  |  |  |  |  | 144 |
|  |  |  |  |  |  |  |  |  | 106 |  |  |  |  | 111 |  |  |  |  |  |  |  |  |  |
|  | 74 |  |  |  |  |  |  |  |  |  |  |  |  |  |  |  |  |  |  |  |  |  | 95 |
|  |  |  |  |  |  |  |  |  |  |  | 60 | 61 |  |  |  |  |  |  |  |  |  |  |  |
|  |  |  | 28 |  |  |  |  |  |  |  |  |  |  |  |  |  |  |  |  |  | 45 |  |  |
|  |  |  |  |  | 6 |  |  |  |  |  |  |  |  |  |  |  |  |  | 19 |  |  |  |  |

Magic Square
Number Fill-ins [B]
Mathematical Formula 24 X 24 + 1 X 12 = 6924

# SHELSEAS MAGIC SQUARE II
## It Works

Positive [B] Start with the top row, right to left in each row, going down.  Begin with the number 1, count each square, only fill-in same shaded squares.

Negative [B] Start with the bottom row, left to right in each row, going up.  Begin with the number 6, count each square, only fill-in same shaded squares.

Objective:  Combine both Positive & Negative number fill-ins [B].  Add each square of numbers in any row, column, or diagonal to proof the same sum of 6924 for a solved Magic Square.  Helpful hints below.

| 1 | 2 | 3 | 4 | 5 | 6 | 7 | 8 | 9 | 10 | 11 | 12 | 13 | 14 | 15 | 16 | 17 | 18 | 19 | 20 | 21 | 22 | 23 | 24 |
|---|---|---|---|---|---|---|---|---|---|---|---|---|---|---|---|---|---|---|---|---|---|---|---|
| 24 | | | | 558 | | | | | | | | | | | | | | | 571 | | | | 1 |
| | 47 | | 532 | | | | | | | | | | | | | | | | | 549 | | 26 | |
| | | 70 | | | | | | | | | 516 | 517 | | | | | | | | | 51 | | |
| | 482 | | 93 | | | | | | | | | | | | | | | | | 76 | | 503 | |
| | | | | 116 | | | | | 466 | | | | | | 471 | | | | 101 | | | | |
| 433 | | | | | | 138 | | | | | | | | | | | 127 | | | | | | 456 |
| | | | | | 163 | 415 | | | | | | | | | | | 426 | 150 | | | | | |
| | | | | | | | | 393 | | 182 | | | 179 | | 400 | | | | | | | | |
| | | | | | | | 368 | | 207 | | | | | 202 | | 377 | | | | | | | |
| | | | | 341 | | | | 232 | | | | | | 225 | | | | | 356 | | | | |
| | | | | | | | 257 | | | 323 | | | 326 | | | 248 | | | | | | | |
| | | 291 | | | | | | | | | 277 | 276 | | | | | | | | | 310 | | |
| | | 267 | | | | | | | | | 301 | 300 | | | | | | | | | 286 | | |
| | | | | | | | 329 | | | 251 | | | 254 | | | 320 | | | | | | | |
| | | | | 221 | | | | 352 | | | | | | | 345 | | | | 236 | | | | |
| | | | | | | | 200 | | 375 | | | | | 370 | | 209 | | | | | | | |
| | | | | | | | | 177 | | 398 | | | 395 | | 184 | | | | | | | | |
| | | | | | 427 | 151 | | | | | | | | | | | 162 | 414 | | | | | |
| 121 | | | | | | 450 | | | | | | | | | | | 439 | | | | | | 144 |
| | | | | 476 | | | | | 106 | | | | | | 111 | | | | 461 | | | | |
| | 74 | | 501 | | | | | | | | | | | | | | | | | 484 | | 95 | |
| | | 526 | | | | | | | | | 60 | 61 | | | | | | | | | 507 | | |
| | 551 | | 28 | | | | | | | | | | | | | | | | | 45 | | 530 | |
| 576 | | | | | 6 | | | | | | | | | | | | | | 19 | | | | 553 |

Magic Square – Number Fill-ins [B] – Mathematical Formula 24 X 24 + 1 X 12 = 6924

# Solved

# [B]

Mathematical Formulas

$8 \times 8 + 1 \times 4 = 260$

$12 \times 12 + 1 \times 6 = 870$

$16 \times 16 + 1 \times 8 = 2056$

$20 \times 20 + 1 \times 10 = 4010$

$24 \times 24 + 1 \times 12 = 6924$

Magic Square
Number Fill-ins [B]
Mathematical Formula

# SHELSEAS MAGIC SQUARE II
## It Works

Positive [B] Diagonals are shown for the same sum of 260

| 8 |    | 6 |    |    | 3 |    | 1 |
|----|----|----|----|----|----|----|----|
|    | 15 |    | 13 | 12 |    | 10 |    |
| 24 |    | 22 |    |    | 19 |    | 17 |
|    | 31 |    | 29 | 28 |    | 26 |    |
|    | 39 |    | 37 | 36 |    | 34 |    |
| 48 |    | 46 |    |    | 43 |    | 41 |
|    | 55 |    | 53 | 52 |    | 50 |    |
| 64 |    | 62 |    |    | 59 |    | 57 |

Negative [B]

|    | 58 |    | 60 | 61 |    | 63 |    |
|----|----|----|----|----|----|----|----|
| 49 |    | 51 |    |    | 54 |    | 56 |
|    | 42 |    | 44 | 45 |    | 47 |    |
| 33 |    | 35 |    |    | 38 |    | 40 |
| 25 |    | 27 |    |    | 30 |    | 32 |
|    | 18 |    | 20 | 21 |    | 23 |    |
| 9 |    | 11 |    |    | 14 |    | 16 |
|    | 2 |    | 4 | 5 |    | 7 |    |

Combination of both Positive & Negative number fill-ins [B] is shown for a solved Magic Square. Objective: Add each square of numbers in any row, column, or diagonal to proof the same sum of 260.

| 8 | 58 | 6 | 60 | 61 | 3 | 63 | 1 |
|----|----|----|----|----|----|----|----|
| 49 | 15 | 51 | 13 | 12 | 54 | 10 | 56 |
| 24 | 42 | 22 | 44 | 45 | 19 | 47 | 17 |
| 33 | 31 | 35 | 29 | 28 | 38 | 26 | 40 |
| 25 | 39 | 27 | 37 | 36 | 30 | 34 | 32 |
| 48 | 18 | 46 | 20 | 21 | 43 | 23 | 41 |
| 9 | 55 | 11 | 53 | 52 | 14 | 50 | 16 |
| 64 | 2 | 62 | 4 | 5 | 59 | 7 | 57 |

Magic Square
Number Fill-ins [B]
Mathematical Formula 8 X 8 + 1 X 4 = 260

# SHELSEAS MAGIC SQUARE II
## It Works

### Positive [B] Diagonals are shown for the same sum of 870

| | | | | | | | | | | | |
|---|---|---|---|---|---|---|---|---|---|---|---|
| 12 | 11 | | | | 7 | 6 | | | | 2 | 1 |
| 24 | 23 | | | 20 | | | 17 | | | 14 | 13 |
| | | 34 | 33 | | 31 | 30 | | 28 | 27 | | |
| | | 46 | 45 | 44 | | | 41 | 40 | 39 | | |
| | 59 | | 57 | 56 | | | 53 | 52 | | 50 | |
| 72 | | 70 | | | 67 | 66 | | | 63 | | 61 |
| 84 | | 82 | | | 79 | 78 | | | 75 | | 73 |
| | 95 | | 93 | 92 | | | 89 | 88 | | 86 | |
| | | 106 | 105 | 104 | | | 101 | 100 | 99 | | |
| | | 118 | 117 | | 115 | 114 | | 112 | 111 | | |
| 132 | 131 | | | 128 | | | 125 | | | 122 | 121 |
| 144 | 143 | | | | 139 | 138 | | | | 134 | 133 |

### Negative [B]

| | | | | | | | | | | | |
|---|---|---|---|---|---|---|---|---|---|---|---|
| | | 135 | 136 | 137 | | | 140 | 141 | 142 | | |
| | | 123 | 124 | | 126 | 127 | | 129 | 130 | | |
| 109 | 110 | | | 113 | | | 116 | | | 119 | 120 |
| 97 | 98 | | | | 102 | 103 | | | | 107 | 108 |
| 85 | | 87 | | | 90 | 91 | | | 94 | | 96 |
| | 74 | | 76 | 77 | | | 80 | 81 | | 83 | |
| | 62 | | 64 | 65 | | | 68 | 69 | | 71 | |
| 49 | | 51 | | | 54 | 55 | | | 58 | | 60 |
| 37 | 38 | | | | 42 | 43 | | | | 47 | 48 |
| 25 | 26 | | | 29 | | | 32 | | | 35 | 36 |
| | | 15 | 16 | | 18 | 19 | | 21 | 22 | | |
| | | 3 | 4 | 5 | | | 8 | 9 | 10 | | |

Combination of both Positive & Negative number fill-ins [B] is shown for a solved Magic Square. Objective: Add each square of numbers in any row, column, or diagonal to proof the same sum of 870.

| | | | | | | | | | | | |
|---|---|---|---|---|---|---|---|---|---|---|---|
| 12 | 11 | 135 | 136 | 137 | 7 | 6 | 140 | 141 | 142 | 2 | 1 |
| 24 | 23 | 123 | 124 | 20 | 126 | 127 | 17 | 129 | 130 | 14 | 13 |
| 109 | 110 | 34 | 33 | 113 | 31 | 30 | 116 | 28 | 27 | 119 | 120 |
| 97 | 98 | 46 | 45 | 44 | 102 | 103 | 41 | 40 | 39 | 107 | 108 |
| 85 | 59 | 87 | 57 | 56 | 90 | 91 | 53 | 52 | 94 | 50 | 96 |
| 72 | 74 | 70 | 76 | 77 | 67 | 66 | 80 | 81 | 63 | 83 | 61 |
| 84 | 62 | 82 | 64 | 65 | 79 | 78 | 68 | 69 | 75 | 71 | 73 |
| 49 | 95 | 51 | 93 | 92 | 54 | 55 | 89 | 88 | 58 | 86 | 60 |
| 37 | 38 | 106 | 105 | 104 | 42 | 43 | 101 | 100 | 99 | 47 | 48 |
| 25 | 26 | 118 | 117 | 29 | 115 | 114 | 32 | 112 | 111 | 35 | 36 |
| 132 | 131 | 15 | 16 | 128 | 18 | 19 | 125 | 21 | 22 | 122 | 121 |
| 144 | 143 | 3 | 4 | 5 | 139 | 138 | 8 | 9 | 10 | 134 | 133 |

Magic Square – Number Fill-ins [B] – Mathematical Formula 12 X 12 + 1 X 6 = 870

# SHELSEAS MAGIC SQUARE II
## It Works

### Positive [B]

|  |  |  |  |  |  |  |  |  |  |  |  |  |  |  |  |
|---|---|---|---|---|---|---|---|---|---|---|---|---|---|---|---|
| 16 | 15 | 14 |  | 12 |  |  |  |  |  |  | 5 |  | 3 | 2 | 1 |
| 32 | 31 |  |  | 28 |  |  | 25 | 24 |  |  | 21 |  |  | 18 | 17 |
| 48 |  |  |  | 44 |  | 42 | 41 | 40 | 39 |  | 37 |  |  |  | 33 |
|  |  |  | 61 | 60 | 59 | 58 |  |  | 55 | 54 | 53 | 52 |  |  |  |
| 80 | 79 | 78 | 77 |  |  |  |  |  |  |  |  | 68 | 67 | 66 | 65 |
|  |  |  | 93 |  | 91 | 90 | 89 | 88 | 87 | 86 |  | 84 |  |  |  |
|  |  | 110 | 109 |  | 107 | 106 |  |  | 103 | 102 |  | 100 | 99 |  |  |
|  | 127 | 126 |  |  | 123 |  | 121 | 120 |  | 118 |  |  | 115 | 114 |  |
|  | 143 | 142 |  |  | 139 |  | 137 | 136 |  | 134 |  |  | 131 | 130 |  |
|  |  | 158 | 157 |  | 155 | 154 |  |  | 151 | 150 |  | 148 | 147 |  |  |
|  |  |  | 173 |  | 171 | 170 | 169 | 168 | 167 | 166 |  | 164 |  |  |  |
| 192 | 191 | 190 | 189 |  |  |  |  |  |  |  |  | 180 | 179 | 178 | 177 |
|  |  |  | 205 | 204 | 203 | 202 |  |  | 199 | 198 | 197 | 196 |  |  |  |
| 224 |  |  |  | 220 |  | 218 | 217 | 216 | 215 |  | 213 |  |  |  | 209 |
| 240 | 239 |  |  | 236 |  |  | 233 | 232 |  |  | 229 |  |  | 226 | 225 |
| 256 | 255 | 254 |  | 252 |  |  |  |  |  |  | 245 |  | 243 | 242 | 241 |

### Negative [B]

|  |  |  |  |  |  |  |  |  |  |  |  |  |  |  |  |
|---|---|---|---|---|---|---|---|---|---|---|---|---|---|---|---|
|  |  |  | 244 |  | 246 | 247 | 248 | 249 | 250 | 251 |  | 253 |  |  |  |
|  |  | 227 | 228 |  | 230 | 231 |  |  | 234 | 235 |  | 237 | 238 |  |  |
|  | 210 | 211 | 212 |  | 214 |  |  |  |  | 219 |  | 221 | 222 | 223 |  |
| 193 | 194 | 195 |  |  |  |  | 200 | 201 |  |  |  |  | 206 | 207 | 208 |
|  |  |  |  | 181 | 182 | 183 | 184 | 185 | 186 | 187 | 188 |  |  |  |  |
| 161 | 162 | 163 |  | 165 |  |  |  |  |  |  | 172 |  | 174 | 175 | 176 |
| 145 | 146 |  |  | 149 |  |  | 152 | 153 |  |  | 156 |  |  | 159 | 160 |
| 129 |  |  | 132 | 133 |  | 135 |  |  | 138 |  | 140 | 141 |  |  | 144 |
| 113 |  |  | 116 | 117 |  | 119 |  |  | 122 |  | 124 | 125 |  |  | 128 |
| 97 | 98 |  |  | 101 |  |  | 104 | 105 |  |  | 108 |  |  | 111 | 112 |
| 81 | 82 | 83 |  | 85 |  |  |  |  |  |  | 92 |  | 94 | 95 | 96 |
|  |  |  |  | 69 | 70 | 71 | 72 | 73 | 74 | 75 | 76 |  |  |  |  |
| 49 | 50 | 51 |  |  |  |  | 56 | 57 |  |  |  |  | 62 | 63 | 64 |
|  | 34 | 35 | 36 |  | 38 |  |  |  |  | 43 |  | 45 | 46 | 47 |  |
|  |  | 19 | 20 |  | 22 | 23 |  |  | 26 | 27 |  | 29 | 30 |  |  |
|  |  |  | 4 |  | 6 | 7 | 8 | 9 | 10 | 11 |  | 13 |  |  |  |

Magic Square
Number Fill-ins [B]
Mathematical Formula 16 X 16 + 1 X 8 = 2056

# SHELSEAS MAGIC SQUARE II
## It Works

Combination of both Positive & Negative number fill-ins [B] is shown for a solved Magic Square. Objective: Add each square of numbers in any row, column, or diagonal to proof the same sum of 2056.

| | | | | | | | | | | | | | | | |
|---|---|---|---|---|---|---|---|---|---|---|---|---|---|---|---|
| 16 | 15 | 14 | 244 | 12 | 246 | 247 | 248 | 249 | 250 | 251 | 5 | 253 | 3 | 2 | 1 |
| 32 | 31 | 227 | 228 | 28 | 230 | 231 | 25 | 24 | 234 | 235 | 21 | 237 | 238 | 18 | 17 |
| 48 | 210 | 211 | 212 | 44 | 214 | 42 | 41 | 40 | 39 | 219 | 37 | 221 | 222 | 223 | 33 |
| 193 | 194 | 195 | 61 | 60 | 59 | 58 | 200 | 201 | 55 | 54 | 53 | 52 | 206 | 207 | 208 |
| 80 | 79 | 78 | 77 | 181 | 182 | 183 | 184 | 185 | 186 | 187 | 188 | 68 | 67 | 66 | 65 |
| 161 | 162 | 163 | 93 | 165 | 91 | 90 | 89 | 88 | 87 | 86 | 172 | 84 | 174 | 175 | 176 |
| 145 | 146 | 110 | 109 | 149 | 107 | 106 | 152 | 153 | 103 | 102 | 156 | 100 | 99 | 159 | 160 |
| 129 | 127 | 126 | 132 | 133 | 123 | 135 | 121 | 120 | 138 | 118 | 140 | 141 | 115 | 114 | 144 |
| 113 | 143 | 142 | 116 | 117 | 139 | 119 | 137 | 136 | 122 | 134 | 124 | 125 | 131 | 130 | 128 |
| 97 | 98 | 158 | 157 | 101 | 155 | 154 | 104 | 105 | 151 | 150 | 108 | 148 | 147 | 111 | 112 |
| 81 | 82 | 83 | 173 | 85 | 171 | 170 | 169 | 168 | 167 | 166 | 92 | 164 | 94 | 95 | 96 |
| 192 | 191 | 190 | 189 | 69 | 70 | 71 | 72 | 73 | 74 | 75 | 76 | 180 | 179 | 178 | 177 |
| 49 | 50 | 51 | 205 | 204 | 203 | 202 | 56 | 57 | 199 | 198 | 197 | 196 | 62 | 63 | 64 |
| 224 | 34 | 35 | 36 | 220 | 38 | 218 | 217 | 216 | 215 | 43 | 213 | 45 | 46 | 47 | 209 |
| 240 | 239 | 19 | 20 | 236 | 22 | 23 | 233 | 232 | 26 | 27 | 229 | 29 | 30 | 226 | 225 |
| 256 | 255 | 254 | 4 | 252 | 6 | 7 | 8 | 9 | 10 | 11 | 245 | 13 | 243 | 242 | 241 |

# 2056

Magic Square
Number Fill-ins [B]
Mathematical Formula 16 X 16 + 1 X 8 = 2056

# SHELSEAS MAGIC SQUARE II
## It Works

### Positive [B]

|    |    |    |    |    |    |    |    |    |    |    |    |    |    |    |    |    |    |    |    |
|----|----|----|----|----|----|----|----|----|----|----|----|----|----|----|----|----|----|----|----|
| 20 | 19 | 18 | 17 |    |    | 14 |    |    |    |    |    |    | 7  |    |    | 4  | 3  | 2  | 1  |
| 40 | 39 | 38 |    |    |    | 34 |    | 32 |    |    | 29 |    | 27 |    |    |    | 23 | 22 | 21 |
| 60 | 59 | 58 | 57 |    |    | 54 |    |    |    |    |    |    | 47 |    |    | 44 | 43 | 42 | 41 |
| 80 |    | 78 |    | 76 |    |    |    | 72 | 71 | 70 | 69 |    |    |    | 65 |    | 63 |    | 61 |
|    |    |    | 97 |    |    | 94 | 93 | 92 | 91 | 90 | 89 | 88 | 87 |    |    | 84 |    |    |    |
|    |    |    |    |    | 115| 114| 113| 112| 111| 110| 109| 108| 107| 106|    |    |    |    |    |
| 140| 139| 138|    | 136| 135|    |    |    |    |    |    |    |    | 126| 125|    | 123| 122| 121|
|    |    |    |    | 156| 155|    | 153| 152| 151| 150| 149| 148|    | 146| 145|    |    |    |    |
|    | 179|    | 177| 176| 175|    | 173|    |    |    |    | 168|    | 166| 165| 164|    | 162|    |
|    |    |    | 197| 196| 195|    | 193|    | 191| 190|    | 188|    | 186| 185| 184|    |    |    |
|    |    |    | 217| 216| 215|    | 213|    | 211| 210|    | 208|    | 206| 205| 204|    |    |    |
|    | 239|    | 237| 236| 235|    | 233|    |    |    |    | 228|    | 226| 225| 224|    | 222|    |
|    |    |    |    | 256| 255|    | 253| 252| 251| 250| 249| 248|    | 246| 245|    |    |    |    |
| 280| 279| 278|    | 276| 275|    |    |    |    |    |    |    |    | 266| 265|    | 263| 262| 261|
|    |    |    |    |    | 295| 294| 293| 292| 291| 290| 289| 288| 287| 286|    |    |    |    |    |
|    |    |    | 317|    |    | 314| 313| 312| 311| 310| 309| 308| 307|    |    | 304|    |    |    |
| 340|    | 338|    | 336|    |    |    | 332| 331| 330| 329|    |    |    | 325|    | 323|    | 321|
| 360| 359| 358| 357|    |    | 354|    |    |    |    |    |    | 347|    |    | 344| 343| 342| 341|
| 380| 379| 378|    |    |    | 374|    | 372|    |    | 369|    | 367|    |    |    | 363| 362| 361|
| 400| 399| 398| 397|    |    | 394|    |    |    |    |    |    | 387|    |    | 384| 383| 382| 381|

### Negative [B]

|    |    |    |    |    |    |    |    |    |    |    |    |    |    |    |    |    |    |    |    |
|----|----|----|----|----|----|----|----|----|----|----|----|----|----|----|----|----|----|----|----|
|    |    |    |    | 385| 386|    | 388| 389| 390| 391| 392| 393|    | 395| 396|    |    |    |    |
|    |    |    | 364| 365| 366|    | 368|    | 370| 371|    | 373|    | 375| 376| 377|    |    |    |
|    |    |    |    | 345| 346|    | 348| 349| 350| 351| 352| 353|    | 355| 356|    |    |    |    |
|    | 322|    | 324|    | 326| 327| 328|    |    |    |    | 333| 334| 335|    | 337|    | 339|    |
| 301| 302| 303|    | 305| 306|    |    |    |    |    |    |    |    | 315| 316|    | 318| 319| 320|
| 281| 282| 283| 284| 285|    |    |    |    |    |    |    |    |    |    | 296| 297| 298| 299| 300|
|    |    |    | 264|    |    | 267| 268| 269| 270| 271| 272| 273| 274|    |    | 277|    |    |    |
| 241| 242| 243| 244|    |    | 247|    |    |    |    |    |    | 254|    |    | 257| 258| 259| 260|
| 221|    | 223|    |    |    | 227|    | 229| 230| 231| 232|    | 234|    |    |    | 238|    | 240|
| 201| 202| 203|    |    |    | 207|    | 209|    |    | 212|    | 214|    |    |    | 218| 219| 220|
| 181| 182| 183|    |    |    | 187|    | 189|    |    | 192|    | 194|    |    |    | 198| 199| 200|
| 161|    | 163|    |    |    | 167|    | 169| 170| 171| 172|    | 174|    |    |    | 178|    | 180|
| 141| 142| 143| 144|    |    | 147|    |    |    |    |    |    | 154|    |    | 157| 158| 159| 160|
|    |    |    | 124|    |    | 127| 128| 129| 130| 131| 132| 133| 134|    |    | 137|    |    |    |
| 101| 102| 103| 104| 105|    |    |    |    |    |    |    |    |    |    | 116| 117| 118| 119| 120|
| 81 | 82 | 83 |    | 85 | 86 |    |    |    |    |    |    |    |    | 95 | 96 |    | 98 | 99 | 100|
|    | 62 |    | 64 |    | 66 | 67 | 68 |    |    |    |    | 73 | 74 | 75 |    | 77 |    | 79 |    |
|    |    |    |    | 45 | 46 |    | 48 | 49 | 50 | 51 | 52 | 53 |    | 55 | 56 |    |    |    |    |
|    |    |    | 24 | 25 | 26 |    | 28 |    | 30 | 31 |    | 33 |    | 35 | 36 | 37 |    |    |    |
|    |    |    |    | 5  | 6  |    | 8  | 9  | 10 | 11 | 12 | 13 |    | 15 | 16 |    |    |    |    |

Magic Square – Number Fill-ins [B] – Mathematical Formula 20 X 20 + 1 X 10 = 4010

# SHELSEAS MAGIC SQUARE II
## It Works

Combination of both Positive & Negative number fill-ins [B] is shown for a solved Magic Square.  Objective:  Add each square of numbers in any row, column, or diagonal to proof the same sum of 4010.

| 20 | 19 | 18 | 17 | 385 | 386 | 14 | 388 | 389 | 390 | 391 | 392 | 393 | 7 | 395 | 396 | 4 | 3 | 2 | 1 |
|---|---|---|---|---|---|---|---|---|---|---|---|---|---|---|---|---|---|---|---|
| 40 | 39 | 38 | 364 | 365 | 366 | 34 | 368 | 32 | 370 | 371 | 29 | 373 | 27 | 375 | 376 | 377 | 23 | 22 | 21 |
| 60 | 59 | 58 | 57 | 345 | 346 | 54 | 348 | 349 | 350 | 351 | 352 | 353 | 47 | 355 | 356 | 44 | 43 | 42 | 41 |
| 80 | 322 | 78 | 324 | 76 | 326 | 327 | 328 | 72 | 71 | 70 | 69 | 333 | 334 | 335 | 65 | 337 | 63 | 339 | 61 |
| 301 | 302 | 303 | 97 | 305 | 306 | 94 | 93 | 92 | 91 | 90 | 89 | 88 | 87 | 315 | 316 | 84 | 318 | 319 | 320 |
| 281 | 282 | 283 | 284 | 285 | 115 | 114 | 113 | 112 | 111 | 110 | 109 | 108 | 107 | 106 | 296 | 297 | 298 | 299 | 300 |
| 140 | 139 | 138 | 264 | 136 | 135 | 267 | 268 | 269 | 270 | 271 | 272 | 273 | 274 | 126 | 125 | 277 | 123 | 122 | 121 |
| 241 | 242 | 243 | 244 | 156 | 155 | 247 | 153 | 152 | 151 | 150 | 149 | 148 | 254 | 146 | 145 | 257 | 258 | 259 | 260 |
| 221 | 179 | 223 | 177 | 176 | 175 | 227 | 173 | 229 | 230 | 231 | 232 | 168 | 234 | 166 | 165 | 164 | 238 | 162 | 240 |
| 201 | 202 | 203 | 197 | 196 | 195 | 207 | 193 | 209 | 191 | 190 | 212 | 188 | 214 | 186 | 185 | 184 | 218 | 219 | 220 |
| 181 | 182 | 183 | 217 | 216 | 215 | 187 | 213 | 189 | 211 | 210 | 192 | 208 | 194 | 206 | 205 | 204 | 198 | 199 | 200 |
| 161 | 239 | 163 | 237 | 236 | 235 | 167 | 233 | 169 | 170 | 171 | 172 | 228 | 174 | 226 | 225 | 224 | 178 | 222 | 180 |
| 141 | 142 | 143 | 144 | 256 | 255 | 147 | 253 | 252 | 251 | 250 | 249 | 248 | 154 | 246 | 245 | 157 | 158 | 159 | 160 |
| 280 | 279 | 278 | 124 | 276 | 275 | 127 | 128 | 129 | 130 | 131 | 132 | 133 | 134 | 266 | 265 | 137 | 263 | 262 | 261 |
| 101 | 102 | 103 | 104 | 105 | 295 | 294 | 293 | 292 | 291 | 290 | 289 | 288 | 287 | 286 | 116 | 117 | 118 | 119 | 120 |
| 81 | 82 | 83 | 317 | 85 | 86 | 314 | 313 | 312 | 311 | 310 | 309 | 308 | 307 | 95 | 96 | 304 | 98 | 99 | 100 |
| 340 | 62 | 338 | 64 | 336 | 66 | 67 | 68 | 332 | 331 | 330 | 329 | 73 | 74 | 75 | 325 | 77 | 323 | 79 | 321 |
| 360 | 359 | 358 | 357 | 45 | 46 | 354 | 48 | 49 | 50 | 51 | 52 | 53 | 347 | 55 | 56 | 344 | 343 | 342 | 341 |
| 380 | 379 | 378 | 24 | 25 | 26 | 374 | 28 | 372 | 30 | 31 | 369 | 33 | 367 | 35 | 36 | 37 | 363 | 362 | 361 |
| 400 | 399 | 398 | 397 | 5 | 6 | 394 | 8 | 9 | 10 | 11 | 12 | 13 | 387 | 15 | 16 | 384 | 383 | 382 | 381 |

# 4010

Magic Square
Number Fill-ins [B]
Mathematical Formula 20 X 20 + 1 X 10 = 4010

# SHELSEAS MAGIC SQUARE II
## It Works

### Positive [B]

|  |  |  |  |  |  |  |  |  |  |  |  |  |  |  |  |  |  |  |  |  |  |  |  |
|---|---|---|---|---|---|---|---|---|---|---|---|---|---|---|---|---|---|---|---|---|---|---|---|
| 24 | 23 | 22 | 21 | 20 |  |  | 17 |  |  |  |  |  |  |  |  | 8 |  |  | 5 | 4 | 3 | 2 | 1 |
| 48 | 47 | 46 |  | 44 |  |  | 41 |  |  |  | 37 | 36 |  |  |  | 32 |  |  | 29 |  | 27 | 26 | 25 |
| 72 | 71 | 70 | 69 | 68 |  |  | 65 |  |  |  |  |  |  |  |  | 56 |  |  | 53 | 52 | 51 | 50 | 49 |
| 96 |  | 94 | 93 | 92 |  |  | 89 |  |  | 86 |  |  | 83 |  |  | 80 |  |  | 77 | 76 | 75 |  | 73 |
| 120 | 119 | 118 | 117 | 116 | 115 |  |  |  |  |  |  |  |  |  |  |  |  | 102 | 101 | 100 | 99 | 98 | 97 |
|  |  |  |  | 140 |  | 138 |  | 136 | 135 | 134 | 133 | 132 | 131 | 130 | 129 |  | 127 |  | 125 |  |  |  |  |
|  |  |  |  |  | 163 |  | 161 | 160 | 159 | 158 | 157 | 156 | 155 | 154 | 153 | 152 |  | 150 |  |  |  |  |  |
| 192 | 191 | 190 | 189 |  |  | 186 |  |  |  | 182 |  |  | 179 |  |  |  | 175 |  |  | 172 | 171 | 170 | 169 |
|  |  |  |  |  | 211 | 210 |  | 208 | 207 | 206 | 205 | 204 | 203 | 202 | 201 |  | 199 | 198 |  |  |  |  |  |
|  |  |  |  |  | 235 | 234 |  | 232 | 231 | 230 | 229 | 228 | 227 | 226 | 225 |  | 223 | 222 |  |  |  |  |  |
|  |  |  | 261 |  | 259 | 258 | 257 | 256 | 255 |  |  |  |  | 250 | 249 | 248 | 247 | 246 |  | 244 |  |  |  |
|  | 287 |  |  |  | 283 | 282 |  | 280 | 279 |  | 277 | 276 |  | 274 | 273 |  | 271 | 270 |  |  |  | 266 |  |
|  | 311 |  |  |  | 307 | 306 |  | 304 | 303 |  | 301 | 300 |  | 298 | 297 |  | 295 | 294 |  |  |  | 290 |  |
|  |  |  | 333 |  | 331 | 330 | 329 | 328 | 327 |  |  |  |  | 322 | 321 | 320 | 319 | 318 |  | 316 |  |  |  |
|  |  |  |  |  | 355 | 354 |  | 352 | 351 | 350 | 349 | 348 | 347 | 346 | 345 |  | 343 | 342 |  |  |  |  |  |
|  |  |  |  |  | 379 | 378 |  | 376 | 375 | 374 | 373 | 372 | 371 | 370 | 369 |  | 367 | 366 |  |  |  |  |  |
| 408 | 407 | 406 | 405 |  |  | 402 |  |  |  | 398 |  |  | 395 |  |  |  | 391 |  |  | 388 | 387 | 386 | 385 |
|  |  |  |  |  | 427 |  | 425 | 424 | 423 | 422 | 421 | 420 | 419 | 418 | 417 | 416 |  | 414 |  |  |  |  |  |
|  |  |  |  | 452 |  | 450 |  | 448 | 447 | 446 | 445 | 444 | 443 | 442 | 441 |  | 439 |  | 437 |  |  |  |  |
| 480 | 479 | 478 | 477 | 476 | 475 |  |  |  |  |  |  |  |  |  |  |  |  | 462 | 461 | 460 | 459 | 458 | 457 |
| 504 |  | 502 | 501 | 500 |  |  | 497 |  |  | 494 |  |  | 491 |  |  | 488 |  |  | 485 | 484 | 483 |  | 481 |
| 528 | 527 | 526 | 525 | 524 |  |  | 521 |  |  |  |  |  |  |  |  | 512 |  |  | 509 | 508 | 507 | 506 | 505 |
| 552 | 551 | 550 |  | 548 |  |  | 545 |  |  |  | 541 | 540 |  |  |  | 536 |  |  | 533 |  | 531 | 530 | 529 |
| 576 | 575 | 574 | 573 | 572 |  |  | 569 |  |  |  |  |  |  |  |  | 560 |  |  | 557 | 556 | 555 | 554 | 553 |

### Negative [B]

|  |  |  |  |  |  |  |  |  |  |  |  |  |  |  |  |  |  |  |  |  |  |  |  |
|---|---|---|---|---|---|---|---|---|---|---|---|---|---|---|---|---|---|---|---|---|---|---|---|
|  |  |  |  |  | 558 | 559 |  | 561 | 562 | 563 | 564 | 565 | 566 | 567 | 568 |  | 570 | 571 |  |  |  |  |  |
|  |  |  | 532 |  | 534 | 535 |  | 537 | 538 | 539 |  |  | 542 | 543 | 544 |  | 546 | 547 |  | 549 |  |  |  |
|  |  |  |  |  | 510 | 511 |  | 513 | 514 | 515 | 516 | 517 | 518 | 519 | 520 |  | 522 | 523 |  |  |  |  |  |
|  | 482 |  |  |  | 486 | 487 |  | 489 | 490 |  | 492 | 493 |  | 495 | 496 |  | 498 | 499 |  |  |  | 503 |  |
|  |  |  |  |  |  | 463 | 464 | 465 | 466 | 467 | 468 | 469 | 470 | 471 | 472 | 473 | 474 |  |  |  |  |  |  |
| 433 | 434 | 435 | 436 |  | 438 |  | 440 |  |  |  |  |  |  |  |  | 449 |  | 451 |  | 453 | 454 | 455 | 456 |
| 409 | 410 | 411 | 412 | 413 |  | 415 |  |  |  |  |  |  |  |  |  |  | 426 |  | 428 | 429 | 430 | 431 | 432 |
|  |  |  |  | 389 | 390 |  | 392 | 393 | 394 |  | 396 | 397 |  | 399 | 400 | 401 |  | 403 | 404 |  |  |  |  |
| 361 | 362 | 363 | 364 | 365 |  |  | 368 |  |  |  |  |  |  |  |  | 377 |  |  | 380 | 381 | 382 | 383 | 384 |
| 337 | 338 | 339 | 340 | 341 |  |  | 344 |  |  |  |  |  |  |  |  | 353 |  |  | 356 | 357 | 358 | 359 | 360 |
| 313 | 314 | 315 |  | 317 |  |  |  |  |  | 323 | 324 | 325 | 326 |  |  |  |  |  | 332 |  | 334 | 335 | 336 |
| 289 |  | 291 | 292 | 293 |  |  | 296 |  |  | 299 |  |  | 302 |  |  | 305 |  |  | 308 | 309 | 310 |  | 312 |
| 265 |  | 267 | 268 | 269 |  |  | 272 |  |  | 275 |  |  | 278 |  |  | 281 |  |  | 284 | 285 | 286 |  | 288 |
| 241 | 242 | 243 |  | 245 |  |  |  |  |  | 251 | 252 | 253 | 254 |  |  |  |  |  | 260 |  | 262 | 263 | 264 |
| 217 | 218 | 219 | 220 | 221 |  |  | 224 |  |  |  |  |  |  |  |  | 233 |  |  | 236 | 237 | 238 | 239 | 240 |
| 193 | 194 | 195 | 196 | 197 |  |  | 200 |  |  |  |  |  |  |  |  | 209 |  |  | 212 | 213 | 214 | 215 | 216 |
|  |  |  |  | 173 | 174 |  | 176 | 177 | 178 |  | 180 | 181 |  | 183 | 184 | 185 |  | 187 | 188 |  |  |  |  |
| 145 | 146 | 147 | 148 | 149 |  | 151 |  |  |  |  |  |  |  |  |  |  | 162 |  | 164 | 165 | 166 | 167 | 168 |
| 121 | 122 | 123 | 124 |  | 126 |  | 128 |  |  |  |  |  |  |  |  | 137 |  | 139 |  | 141 | 142 | 143 | 144 |
|  |  |  |  |  |  | 103 | 104 | 105 | 106 | 107 | 108 | 109 | 110 | 111 | 112 | 113 | 114 |  |  |  |  |  |  |
|  | 74 |  |  |  | 78 | 79 |  | 81 | 82 |  | 84 | 85 |  | 87 | 88 |  | 90 | 91 |  |  |  | 95 |  |
|  |  |  |  |  | 54 | 55 |  | 57 | 58 | 59 | 60 | 61 | 62 | 63 | 64 |  | 66 | 67 |  |  |  |  |  |
|  |  |  | 28 |  | 30 | 31 |  | 33 | 34 | 35 |  |  | 38 | 39 | 40 |  | 42 | 43 |  | 45 |  |  |  |
|  |  |  |  |  | 6 | 7 |  | 9 | 10 | 11 | 12 | 13 | 14 | 15 | 16 |  | 18 | 19 |  |  |  |  |  |

Magic Square – Number Fill-ins [B] – Mathematical Formula 24 X 24 + 1 X 12 = 6924

Combination of both Positive & Negative number fill-ins [B] is shown for a solved
Magic Square.  Objective:  Add each square of numbers in any row, column, or diagonal
to proof the same sum of 6924.

| 24 | 23 | 22 | 21 | 20 | 558 | 559 | 17 | 561 | 562 | 563 | 564 | 565 | 566 | 567 | 568 | 8 | 570 | 571 | 5 | 4 | 3 | 2 | 1 |
|---|---|---|---|---|---|---|---|---|---|---|---|---|---|---|---|---|---|---|---|---|---|---|---|
| 48 | 47 | 46 | 532 | 44 | 534 | 535 | 41 | 537 | 538 | 539 | 37 | 36 | 542 | 543 | 544 | 32 | 546 | 547 | 29 | 549 | 27 | 26 | 25 |
| 72 | 71 | 70 | 69 | 68 | 510 | 511 | 65 | 513 | 514 | 515 | 516 | 517 | 518 | 519 | 520 | 56 | 522 | 523 | 53 | 52 | 51 | 50 | 49 |
| 96 | 482 | 94 | 93 | 92 | 486 | 487 | 89 | 489 | 490 | 86 | 492 | 493 | 83 | 495 | 496 | 80 | 498 | 499 | 77 | 76 | 75 | 503 | 73 |
| 120 | 119 | 118 | 117 | 116 | 115 | 463 | 464 | 465 | 466 | 467 | 468 | 469 | 470 | 471 | 472 | 473 | 474 | 102 | 101 | 100 | 99 | 98 | 97 |
| 433 | 434 | 435 | 436 | 140 | 438 | 138 | 440 | 136 | 135 | 134 | 133 | 132 | 131 | 130 | 129 | 449 | 127 | 451 | 125 | 453 | 454 | 455 | 456 |
| 409 | 410 | 411 | 412 | 413 | 163 | 415 | 161 | 160 | 159 | 158 | 157 | 156 | 155 | 154 | 153 | 152 | 426 | 150 | 428 | 429 | 430 | 431 | 432 |
| 192 | 191 | 190 | 189 | 389 | 390 | 186 | 392 | 393 | 394 | 182 | 396 | 397 | 179 | 399 | 400 | 401 | 175 | 403 | 404 | 172 | 171 | 170 | 169 |
| 361 | 362 | 363 | 364 | 365 | 211 | 210 | 368 | 208 | 207 | 206 | 205 | 204 | 203 | 202 | 201 | 377 | 199 | 198 | 380 | 381 | 382 | 383 | 384 |
| 337 | 338 | 339 | 340 | 341 | 235 | 234 | 344 | 232 | 231 | 230 | 229 | 228 | 227 | 226 | 225 | 353 | 223 | 222 | 356 | 357 | 358 | 359 | 360 |
| 313 | 314 | 315 | 261 | 317 | 259 | 258 | 257 | 256 | 255 | 323 | 324 | 325 | 326 | 250 | 249 | 248 | 247 | 246 | 332 | 244 | 334 | 335 | 336 |
| 289 | 287 | 291 | 292 | 293 | 283 | 282 | 296 | 280 | 279 | 299 | 277 | 276 | 302 | 274 | 273 | 305 | 271 | 270 | 308 | 309 | 310 | 266 | 312 |
| 265 | 311 | 267 | 268 | 269 | 307 | 306 | 272 | 304 | 303 | 275 | 301 | 300 | 278 | 298 | 297 | 281 | 295 | 294 | 284 | 285 | 286 | 290 | 288 |
| 241 | 242 | 243 | 333 | 245 | 331 | 330 | 329 | 328 | 327 | 251 | 252 | 253 | 254 | 322 | 321 | 320 | 319 | 318 | 260 | 316 | 262 | 263 | 264 |
| 217 | 218 | 219 | 220 | 221 | 355 | 354 | 224 | 352 | 351 | 350 | 349 | 348 | 347 | 346 | 345 | 233 | 343 | 342 | 236 | 237 | 238 | 239 | 240 |
| 193 | 194 | 195 | 196 | 197 | 379 | 378 | 200 | 376 | 375 | 374 | 373 | 372 | 371 | 370 | 369 | 209 | 367 | 366 | 212 | 213 | 214 | 215 | 216 |
| 408 | 407 | 406 | 405 | 173 | 174 | 402 | 176 | 177 | 178 | 398 | 180 | 181 | 395 | 183 | 184 | 185 | 391 | 187 | 188 | 388 | 387 | 386 | 385 |
| 145 | 146 | 147 | 148 | 149 | 427 | 151 | 425 | 424 | 423 | 422 | 421 | 420 | 419 | 418 | 417 | 416 | 162 | 414 | 164 | 165 | 166 | 167 | 168 |
| 121 | 122 | 123 | 124 | 452 | 126 | 450 | 128 | 448 | 447 | 446 | 445 | 444 | 443 | 442 | 441 | 137 | 439 | 139 | 437 | 141 | 142 | 143 | 144 |
| 480 | 479 | 478 | 477 | 476 | 475 | 103 | 104 | 105 | 106 | 107 | 108 | 109 | 110 | 111 | 112 | 113 | 114 | 462 | 461 | 460 | 459 | 458 | 457 |
| 504 | 74 | 502 | 501 | 500 | 78 | 79 | 497 | 81 | 82 | 494 | 84 | 85 | 491 | 87 | 88 | 488 | 90 | 91 | 485 | 484 | 483 | 95 | 481 |
| 528 | 527 | 526 | 525 | 524 | 54 | 55 | 521 | 57 | 58 | 59 | 60 | 61 | 62 | 63 | 64 | 512 | 66 | 67 | 509 | 508 | 507 | 506 | 505 |
| 552 | 551 | 550 | 28 | 548 | 30 | 31 | 545 | 33 | 34 | 35 | 541 | 540 | 38 | 39 | 40 | 536 | 42 | 43 | 533 | 45 | 531 | 530 | 529 |
| 576 | 575 | 574 | 573 | 572 | 6 | 7 | 569 | 9 | 10 | 11 | 12 | 13 | 14 | 15 | 16 | 560 | 18 | 19 | 557 | 556 | 555 | 554 | 553 |

# 6924

Magic Square
Number Fill-ins [B]
Mathematical Formula 24 X 24 + 1 X 12 = 6924

SHELSEAS MAGIC SQUARE II
It Works

The Number Fill-ins [C] Will Be Your Guideline
For A Solved Double Even – Magic Square.

# Number
# Fill-ins

# [C]

Positive [C] Start with the bottom row, right to left in each row, going up.  Begin with the number shown hint, count each square, only fill-in same shaded squares.

Negative [C] Start with the top row, left to right in each row, going down.  Begin with the number shown hint, count each square, only fill-in same shaded squares.

Objective:  Combine both Positive & Negative number fill-ins [C] in each Magic Square.  Add each square of numbers in any row, column, or diagonal to proof the same sum of integer for a solved Magic Square.

Mathematical Formulas

$8 \times 8 + 1 \times 4 = 260$

$12 \times 12 + 1 \times 6 = 870$

$16 \times 16 + 1 \times 8 = 2056$

$20 \times 20 + 1 \times 10 = 4010$

$24 \times 24 + 1 \times 12 = 6924$

# SHELSEAS MAGIC SQUARE II
## It Works

Positive [C] Start with the bottom row, right to left in each row, going up. Begin with the number 1, count each square, only fill-in shaded squares.

| 16 |    |    | 13 |
|----|----|----|----|
|    | 11 | 10 |    |
|    | 7  | 6  |    |
| 4  |    |    | 1  |

Negative [C] Start with the top row, left to right in each row, going down.  Begin with the number 2, count each square, only fill-in shaded squares.

|    | 2  | 3  |    |
|----|----|----|----|
| 5  |    |    | 8  |
| 9  |    |    | 12 |
|    | 14 | 15 |    |

Combination of both Positive & Negative number fill-ins [C] is shown for a solved Magic Square.  Objective:  Add each square of numbers in any row, column or diagonal to proof the same sum of 34.

| 16 | 2  | 3  | 13 |
|----|----|----|----|
| 5  | 11 | 10 | 8  |
| 9  | 7  | 6  | 12 |
| 4  | 14 | 15 | 1  |

Magic Square
Number Fill-ins [C]
Mathematical Formula 4 X 4 + 1 X 2 = 34

# SHELSEAS MAGIC SQUARE II
## It Works

Positive [C] Start with the bottom row, right to left in each row, going up. Begin with the number 1, count each square, only fill-in shaded squares. Helpful hints below.

| 64 |    |    |    |    |    |    | 57 |
|----|----|----|----|----|----|----|----|
|    | 55 |    |    |    |    | 50 |    |
|    |    | 46 |    |    | 43 |    |    |
|    |    |    | 37 | 36 |    |    |    |
|    |    |    | 29 | 28 |    |    |    |
|    |    | 22 |    |    | 19 |    |    |
|    | 15 |    |    |    |    | 10 |    |
| 8  |    |    |    |    |    |    | 1  |

Negative [C] Start with the top row, left to right in each row, going down. Begin with the number 2, count each square, only fill-in shaded squares. Helpful hints below.

|    | 2  |    |    |    |    | 7  |    |
|----|----|----|----|----|----|----|----|
|    |    |    | 12 | 13 |    |    |    |
| 17 |    |    |    |    |    |    | 24 |
|    |    | 27 |    |    | 30 |    |    |
|    |    | 35 |    |    | 38 |    |    |
| 41 |    |    |    |    |    |    | 48 |
|    |    |    | 52 | 53 |    |    |    |
|    | 58 |    |    |    |    | 63 |    |

Objective: Combine both Positive & Negative number fill-ins [C]. Add each square of numbers in any row, column, or diagonal to proof the same sum of 260 for a solved Magic Square. Helpful hints below.

| 64 | 2  |    |    |    |    | 7  | 57 |
|----|----|----|----|----|----|----|----|
|    | 55 |    | 12 | 13 |    | 50 |    |
| 17 |    | 46 |    |    | 43 |    | 24 |
|    |    | 27 | 37 | 36 | 30 |    |    |
|    |    | 35 | 29 | 28 | 38 |    |    |
| 41 |    | 22 |    |    | 19 |    | 48 |
|    | 15 |    | 52 | 53 |    | 10 |    |
| 8  | 58 |    |    |    |    | 63 | 1  |

Magic Square – Number Fill-ins [C] – Mathematical Formula 8 X 8 + 1 X 4 = 260

# SHELSEAS MAGIC SQUARE II
## It Works

Positive [C] Start with the bottom row, right to left in each row, going up. Begin with the number 1, count each square, only fill-in shaded squares. Helpful hints below.

| | | | | | | | | | | | |
|---|---|---|---|---|---|---|---|---|---|---|---|
| 144 | | | | | | | | | | | 133 |
| | | | 129 | | | | | 124 | | | |
| | | 118 | | | | | | | 111 | | |
| | 107 | | | | | | | | | 98 | |
| | | | | 92 | | | 89 | | | | |
| | | | | | 79 | 78 | | | | | |
| | | | | | 67 | 66 | | | | | |
| | | | | 56 | | | 53 | | | | |
| | 47 | | | | | | | | | 38 | |
| | | 34 | | | | | | | 27 | | |
| | | | 21 | | | | | 16 | | | |
| 12 | | | | | | | | | | | 1 |

Negative [C] Start with the top row, left to right in each row, going down. Begin with the number 4, count each square, only fill-in shaded squares. Helpful hints below.

| | | | | | | | | | | | |
|---|---|---|---|---|---|---|---|---|---|---|---|
| | | | 4 | | | | | 9 | | | |
| | | | | 18 | 19 | | | | | | |
| | | | 29 | | | 32 | | | | | |
| 37 | | | | | | | | | | | 48 |
| | | 51 | | | | | | | 58 | | |
| | 62 | | | | | | | | | 71 | |
| | 74 | | | | | | | | | 83 | |
| | | 87 | | | | | | | 94 | | |
| 97 | | | | | | | | | | | 108 |
| | | | 113 | | | 116 | | | | | |
| | | | | 126 | 127 | | | | | | |
| | | | 136 | | | | | 141 | | | |

Magic Square
Number Fill-ins [C]
Mathematical Formula 12 X 12 + 1 X 6 = 870

# SHELSEAS MAGIC SQUARE II
## It Works

Positive [C] Start with the bottom row, right to left in each row, going up. Begin with the number 1, count each square, only fill-in same shaded squares.

Negative [C] Start with the top row, left to right in each row, going down. Begin with the number 4, count each square, only fill-in same shaded squares.

Objective: Combine both Positive & Negative number fill-ins [C]. Add each square of numbers in any row, column, or diagonal to proof the same sum of 870 for a solved Magic Square. Helpful hints below.

| | | | | | | | | | | | |
|---|---|---|---|---|---|---|---|---|---|---|---|
| 144 | | | 4 | | | | | 9 | | | 133 |
| | | | 129 | | 18 | 19 | | 124 | | | |
| | | 118 | | 29 | | | 32 | | 111 | | |
| 37 | 107 | | | | | | | | | 98 | 48 |
| | | 51 | | 92 | | | 89 | | 58 | | |
| | 62 | | | | 79 | 78 | | | | 71 | |
| | 74 | | | | 67 | 66 | | | | 83 | |
| | | 87 | | 56 | | | 53 | | 94 | | |
| 97 | 47 | | | | | | | | | 38 | 108 |
| | | 34 | | 113 | | | 116 | | 27 | | |
| | | | 21 | | 126 | 127 | | 16 | | | |
| 12 | | | 136 | | | | | 141 | | | 1 |

Magic Square
Number Fill-ins [C]
Mathematical Formula 12 X 12 + 1 X 6 = 870

# SHELSEAS MAGIC SQUARE II
## It Works

Positive [C] Start with the bottom row, right to left in each row, going up. Begin with the number 1, count each square, only fill-in shaded squares. Diagonals are shown for the same sum of 2056. Helpful hints below.

| | | | | | | | | | | | | | | | |
|---|---|---|---|---|---|---|---|---|---|---|---|---|---|---|---|
| 256 | | | | | | | | | | | | | | | 241 |
| | 239 | | | | | | | | | | | | | 226 | |
| | | 222 | | | | | | | | | | | 211 | | |
| | | | 205 | | | | | | | | | 196 | | | |
| | | | | 188 | | | | | | | 181 | | | | |
| | | | | | 171 | | | | | 166 | | | | | |
| | | | | | | 154 | | | 151 | | | | | | |
| | | | | | | | 137 | 136 | | | | | | | |
| | | | | | | | 121 | 120 | | | | | | | |
| | | | | | | 106 | | | 103 | | | | | | |
| | | | | | 91 | | | | | 86 | | | | | |
| | | | | 76 | | | | | | | 69 | | | | |
| | | | 61 | | | | | | | | | 52 | | | |
| | | 46 | | | | | | | | | | | 35 | | |
| | 31 | | | | | | | | | | | | | 18 | |
| 16 | | | | | | | | | | | | | | | 1 |

Negative [C] Start with the top row, left to right in each row, going down. Begin with the number 5, count each square, only fill-in shaded squares. Helpful hints below.

| | | | | | | | | | | | | | | | |
|---|---|---|---|---|---|---|---|---|---|---|---|---|---|---|---|
| | | | | 5 | | | | | | | 12 | | | | |
| | | 20 | | | | | | | | | | | 29 | | |
| | | | | | | 40 | 41 | | | | | | | | |
| | 50 | | | | | | | | | | | | | 63 | |
| 65 | | | | | | | | | | | | | | | 80 |
| | | | | | 87 | | | | 90 | | | | | | |
| | | | | 102 | | | | | | 107 | | | | | |
| | | 115 | | | | | | | | | | | 126 | | |
| | | 131 | | | | | | | | | | | 142 | | |
| | | | | 150 | | | | | | 155 | | | | | |
| | | | | | 167 | | | | 170 | | | | | | |
| 177 | | | | | | | | | | | | | | | 192 |
| | 194 | | | | | | | | | | | | | 207 | |
| | | | | | | 216 | 217 | | | | | | | | |
| | | 228 | | | | | | | | | | | 237 | | |
| | | | | 245 | | | | | | | 252 | | | | |

Magic Square – Number Fill-ins [C] – Mathematical Formula 16 X 16 + 1 X 8 = 2056

## SHELSEAS MAGIC SQUARE II
### It Works

Positive [C] Start with the bottom row, right to left in each row, going up. Begin with the number 1, count each square, only fill-in same shaded squares. Diagonals are shown for the same sum of 2056.

Negative [C] Start with the top row, left to right in each row, going down. Begin with the number 5, count each square, only fill-in same shaded squares.

Objective: Combine both Positive & Negative number fill-ins [C]. Add each square of numbers in any row, column, or diagonal to proof the same sum of 2056 for a solved Magic Square. Helpful hints below.

|  |  |  |  |  |  |  |  |  |  |  |  |  |  |  |  |
|---|---|---|---|---|---|---|---|---|---|---|---|---|---|---|---|
| 256 |  |  |  | 5 |  |  |  |  |  |  | 12 |  |  |  | 241 |
|  | 239 |  | 20 |  |  |  |  |  |  |  |  | 29 |  | 226 |  |
|  |  | 222 |  |  |  |  | 40 | 41 |  |  |  |  | 211 |  |  |
|  | 50 |  | 205 |  |  |  |  |  |  |  |  | 196 |  | 63 |  |
| 65 |  |  |  | 188 |  |  |  |  |  |  | 181 |  |  |  | 80 |
|  |  |  |  |  | 171 | 87 |  |  | 90 | 166 |  |  |  |  |  |
|  |  |  |  |  | 102 | 154 |  |  | 151 | 107 |  |  |  |  |  |
|  |  | 115 |  |  |  |  | 137 | 136 |  |  |  |  | 126 |  |  |
|  |  | 131 |  |  |  |  | 121 | 120 |  |  |  |  | 142 |  |  |
|  |  |  |  |  | 150 | 106 |  |  | 103 | 155 |  |  |  |  |  |
|  |  |  |  |  | 91 | 167 |  |  | 170 | 86 |  |  |  |  |  |
| 177 |  |  |  | 76 |  |  |  |  |  |  | 69 |  |  |  | 192 |
|  | 194 |  | 61 |  |  |  |  |  |  |  |  | 52 |  | 207 |  |
|  |  | 46 |  |  |  |  | 216 | 217 |  |  |  |  | 35 |  |  |
|  | 31 |  | 228 |  |  |  |  |  |  |  |  | 237 |  | 18 |  |
| 16 |  |  |  | 245 |  |  |  |  |  |  | 252 |  |  |  | 1 |

Magic Square
Number Fill-ins [C]
Mathematical Formula 16 X 16 + 1 X 8 = 2056

# SHELSEAS MAGIC SQUARE II
## It Works

Positive [C] Start with the bottom row, right to left in each row, going up. Begin with the number 1, count each square, only fill-in shaded squares. Helpful hints below.

| | | | | | | | | | | | | | | | | | | | |
|---|---|---|---|---|---|---|---|---|---|---|---|---|---|---|---|---|---|---|---|
| 400 | | | | | | | | | | | | | | | | | | | 381 |
| | | | | | | | | 372 | | | 369 | | | | | | | | |
| | | | 357 | | | | | | | | | | | | | 344 | | | |
| | | 338 | | | | | | | | | | | | | | | 323 | | |
| | | | | 316 | | | | | | | | | | | 305 | | | | |
| | | | | | | | | | 291 | 290 | | | | | | | | | |
| | | | | | | | 273 | | | | | 268 | | | | | | | |
| | | | | | | 254 | | | | | | | 247 | | | | | | |
| | 239 | | | | | | | | | | | | | | | | | 222 | |
| | | | | | 215 | | | | | | | | | 206 | | | | | |
| | | | | | 195 | | | | | | | | | 186 | | | | | |
| | 179 | | | | | | | | | | | | | | | | | 162 | |
| | | | | | | 154 | | | | | | | 147 | | | | | | |
| | | | | | | | 133 | | | | | 128 | | | | | | | |
| | | | | | | | | | 111 | 110 | | | | | | | | | |
| | | | | 96 | | | | | | | | | | | 85 | | | | |
| | | 78 | | | | | | | | | | | | | | | 63 | | |
| | | | 57 | | | | | | | | | | | | | 44 | | | |
| | | | | | | | | 32 | | | 29 | | | | | | | | |
| 20 | | | | | | | | | | | | | | | | | | | 1 |

Magic Square
Number Fill-ins [C]
Mathematical Formula 20 X 20 + 1 X 10 = 4010

# SHELSEAS MAGIC SQUARE II
## It Works

Negative [C] Start with the top row, left to right in each row, going down.  Begin with the number 6, count each square, only fill-in shaded squares.  Helpful hints below.

| 1 | 2 | 3 | 4 | 5 | 6 | 7 | 8 | 9 | 10 | 11 | 12 | 13 | 14 | 15 | 16 | 17 | 18 | 19 | 20 |
|---|---|---|---|---|---|---|---|---|----|----|----|----|----|----|----|----|----|----|----|
|  |  |  |  |  | 6 |  |  |  |  |  |  |  |  | 15 |  |  |  |  |  |
|  |  |  |  |  |  |  |  |  | 30 | 31 |  |  |  |  |  |  |  |  |  |
|  |  |  |  |  |  |  | 48 |  |  |  |  | 53 |  |  |  |  |  |  |  |
|  |  |  |  |  |  | 67 |  |  |  |  |  |  | 74 |  |  |  |  |  |  |
|  |  |  |  |  |  |  |  | 89 |  |  | 92 |  |  |  |  |  |  |  |  |
| 101 |  |  |  |  |  |  |  |  |  |  |  |  |  |  |  |  |  |  | 120 |
|  |  |  | 124 |  |  |  |  |  |  |  |  |  |  |  |  | 137 |  |  |  |
|  |  | 143 |  |  |  |  |  |  |  |  |  |  |  |  |  |  | 158 |  |  |
|  |  |  |  | 165 |  |  |  |  |  |  |  |  |  |  | 176 |  |  |  |  |
|  | 182 |  |  |  |  |  |  |  |  |  |  |  |  |  |  |  |  | 199 |  |
|  | 202 |  |  |  |  |  |  |  |  |  |  |  |  |  |  |  |  | 219 |  |
|  |  |  |  | 225 |  |  |  |  |  |  |  |  |  |  | 236 |  |  |  |  |
|  |  | 243 |  |  |  |  |  |  |  |  |  |  |  |  |  |  | 258 |  |  |
|  |  |  | 264 |  |  |  |  |  |  |  |  |  |  |  |  | 277 |  |  |  |
| 281 |  |  |  |  |  |  |  |  |  |  |  |  |  |  |  |  |  |  | 300 |
|  |  |  |  |  |  |  |  | 309 |  |  | 312 |  |  |  |  |  |  |  |  |
|  |  |  |  |  |  | 327 |  |  |  |  |  |  | 334 |  |  |  |  |  |  |
|  |  |  |  |  |  |  | 348 |  |  |  |  | 353 |  |  |  |  |  |  |  |
|  |  |  |  |  |  |  |  |  | 370 | 371 |  |  |  |  |  |  |  |  |  |
|  |  |  |  |  | 386 |  |  |  |  |  |  | 395 |  |  |  |  |  |  |  |

Magic Square
Number Fill-ins [C]
Mathematical Formula 20 X 20 + 1 X 10 = 4010

# SHELSEAS MAGIC SQUARE II
## It Works

Positive [C] Start with the bottom row, right to left in each row, going up.  Begin with the number 1, count each square, only fill-in same shaded squares.

Negative [C] Start with the top row, left to right in each row, going down.  Begin with the number 6, count each square, only fill-in same shaded squares.

Objective:  Combine both Positive & Negative number fill-ins [C].  Add each square of numbers in any row, column, or diagonal to proof the same sum of 4010 for a solved Magic Square.  Helpful hints below.

| | | | | | | | | | | | | | | | | | | | |
|---|---|---|---|---|---|---|---|---|---|---|---|---|---|---|---|---|---|---|---|
| 400 | | | | 6 | | | | | | | | | | | 15 | | | | 381 |
| | | | | | | | 372 | 30 | 31 | 369 | | | | | | | | | |
| | | 357 | | | 48 | | | | | | 53 | | | | | 344 | | | |
| | 338 | | | | 67 | | | | | | | 74 | | | | | 323 | | |
| | | | 316 | | | 89 | | | 92 | | | | | 305 | | | | | |
| 101 | | | | | | | 291 | 290 | | | | | | | | | | | 120 |
| | | 124 | | | 273 | | | | | | 268 | | | | | 137 | | | |
| | | 143 | | | 254 | | | | | | 247 | | | | | 158 | | | |
| | 239 | | 165 | | | | | | | | | | | 176 | | | | 222 | |
| | 182 | | | 215 | | | | | | | | | 206 | | | | | 199 | |
| | 202 | | | 195 | | | | | | | | | 186 | | | | | 219 | |
| | 179 | | 225 | | | | | | | | | | | 236 | | | | 162 | |
| | | 243 | | | 154 | | | | | | 147 | | | | | 258 | | | |
| | | 264 | | | 133 | | | | | | 128 | | | | | 277 | | | |
| 281 | | | | | | | 111 | 110 | | | | | | | | | | | 300 |
| | | | 96 | | | 309 | | | 312 | | | | | 85 | | | | | |
| | | 78 | | | 327 | | | | | | | 334 | | | | | 63 | | |
| | | 57 | | | 348 | | | | | | 353 | | | | | 44 | | | |
| | | | | | | | 32 | 370 | 371 | 29 | | | | | | | | | |
| 20 | | | | 386 | | | | | | | | | | | 395 | | | | 1 |

Magic Square – Number Fill-ins [C] – Mathematical Formula 20 X 20 + 1 X 10 = 4010

# SHELSEAS MAGIC SQUARE II
## It Works

Positive [C] Start with the bottom row, right to left in each row, going up. Begin with the number 1, count each square, only fill-in shaded squares. Helpful hints below.

| 1 | 2 | 3 | 4 | 5 | 6 | 7 | 8 | 9 | 10 | 11 | 12 | 13 | 14 | 15 | 16 | 17 | 18 | 19 | 20 | 21 | 22 | 23 | 24 |
|---|---|---|---|---|---|---|---|---|----|----|----|----|----|----|----|----|----|----|----|----|----|----|----|
| 576 |  |  |  |  |  |  |  |  |  |  |  |  |  |  |  |  |  |  |  |  |  |  | 553 |
|  | 551 |  |  |  |  |  |  |  |  |  |  |  |  |  |  |  |  |  |  |  |  | 530 |  |
|  |  | 526 |  |  |  |  |  |  |  |  |  |  |  |  |  |  |  |  |  |  | 507 |  |  |
|  |  |  | 501 |  |  |  |  |  |  |  |  |  |  |  |  |  |  |  |  | 484 |  |  |  |
|  |  |  |  | 476 |  |  |  |  |  |  |  |  |  |  |  |  |  |  | 461 |  |  |  |  |
|  |  |  |  |  |  |  |  |  | 446 |  |  | 443 |  |  |  |  |  |  |  |  |  |  |  |
|  |  |  |  |  |  |  | 425 |  |  |  |  |  |  |  | 416 |  |  |  |  |  |  |  |  |
|  |  |  |  |  |  | 402 |  |  |  |  |  |  |  |  |  | 391 |  |  |  |  |  |  |  |
|  |  |  |  |  |  |  |  | 376 |  |  |  |  |  | 369 |  |  |  |  |  |  |  |  |  |
|  |  |  |  |  |  |  |  |  | 351 |  |  |  | 346 |  |  |  |  |  |  |  |  |  |  |
|  |  |  |  |  | 331 |  |  |  |  |  |  |  |  |  |  |  | 318 |  |  |  |  |  |  |
|  |  |  |  |  |  |  |  |  |  |  | 301 | 300 |  |  |  |  |  |  |  |  |  |  |  |
|  |  |  |  |  |  |  |  |  |  |  | 277 | 276 |  |  |  |  |  |  |  |  |  |  |  |
|  |  |  |  |  | 259 |  |  |  |  |  |  |  |  |  |  |  | 246 |  |  |  |  |  |  |
|  |  |  |  |  |  |  |  |  | 231 |  |  |  | 226 |  |  |  |  |  |  |  |  |  |  |
|  |  |  |  |  |  |  |  | 208 |  |  |  |  |  | 201 |  |  |  |  |  |  |  |  |  |
|  |  |  |  |  |  | 186 |  |  |  |  |  |  |  |  |  | 175 |  |  |  |  |  |  |  |
|  |  |  |  |  |  |  | 161 |  |  |  |  |  |  |  | 152 |  |  |  |  |  |  |  |  |
|  |  |  |  |  |  |  |  |  | 134 |  |  | 131 |  |  |  |  |  |  |  |  |  |  |  |
|  |  |  |  | 116 |  |  |  |  |  |  |  |  |  |  |  |  |  |  | 101 |  |  |  |  |
|  |  |  | 93 |  |  |  |  |  |  |  |  |  |  |  |  |  |  |  |  | 76 |  |  |  |
|  |  | 70 |  |  |  |  |  |  |  |  |  |  |  |  |  |  |  |  |  |  | 51 |  |  |
|  | 47 |  |  |  |  |  |  |  |  |  |  |  |  |  |  |  |  |  |  |  |  | 26 |  |
| 24 |  |  |  |  |  |  |  |  |  |  |  |  |  |  |  |  |  |  |  |  |  |  | 1 |

Magic Square
Number Fill-ins [C]
Mathematical Formula 24 X 24 + 1 X 12 = 6924

# SHELSEAS MAGIC SQUARE II
## It Works

Negative [C] Start with the top row, left to right in each row, going down. Begin with the number 5, count each square, only fill-in shaded squares. Helpful hints below.

| 1 | 2 | 3 | 4 | 5 | 6 | 7 | 8 | 9 | 10 | 11 | 12 | 13 | 14 | 15 | 16 | 17 | 18 | 19 | 20 | 21 | 22 | 23 | 24 |
|---|---|---|---|---|---|---|---|---|----|----|----|----|----|----|----|----|----|----|----|----|----|----|----|
| | | | | 5 | | | | | | | | | | | | | | | | 20 | | | |
| | | | 28 | | | | | | | | | | | | | | | | | | 45 | | |
| | | | | | | | | | | 59 | | | 62 | | | | | | | | | | |
| | 74 | | | | | | | | | | | | | | | | | | | | | 95 | |
| 97 | | | | | | | | | | | | | | | | | | | | | | | 120 |
| | | | | | 126 | | | | | | | | | | | | | 139 | | | | | |
| | | | | | | 151 | | | | | | | | | | | 162 | | | | | | |
| | | | | | | | | | 178 | | | | | 183 | | | | | | | | | |
| | | | | | | | | | | | 204 | 205 | | | | | | | | | | | |
| | | | | | | | 224 | | | | | | | | | 233 | | | | | | | |
| | | 243 | | | | | | | | | | | | | | | | | | | 262 | | |
| | | | | | | | | 273 | | | | | | | 280 | | | | | | | | |
| | | | | | | | | 297 | | | | | | | 304 | | | | | | | | |
| | | 315 | | | | | | | | | | | | | | | | | | | 334 | | |
| | | | | | | | 344 | | | | | | | | | 353 | | | | | | | |
| | | | | | | | | | | | 372 | 373 | | | | | | | | | | | |
| | | | | | | | | | 394 | | | | | 399 | | | | | | | | | |
| | | | | | | 415 | | | | | | | | | | | 426 | | | | | | |
| | | | | | 438 | | | | | | | | | | | | | 451 | | | | | |
| 457 | | | | | | | | | | | | | | | | | | | | | | | 480 |
| | 482 | | | | | | | | | | | | | | | | | | | | | 503 | |
| | | | | | | | | | | 515 | | | 518 | | | | | | | | | | |
| | | | 532 | | | | | | | | | | | | | | | | | | 549 | | |
| | | | | 557 | | | | | | | | | | | | | | | | 572 | | | |

Magic Square
Number Fill-ins [C]
Mathematical Formula 24 X 24 + 1 X 12 = 6924

# SHELSEAS MAGIC SQUARE II
## It Works

Positive [C] Start with the bottom row, right to left in each row, going up.  Begin with the number 1, count each square, only fill-in same shaded squares.

Negative [C] Start with the top row, left to right in each row, going down.  Begin with the number 5, count each square, only fill-in same shaded squares.

Objective:  Combine both Positive & Negative number fill-ins [C].  Add each square of numbers in any row, column, or diagonal to proof the same sum of 6924 for a solved Magic Square.  Helpful hints below.

| 1 | 2 | 3 | 4 | 5 | 6 | 7 | 8 | 9 | 10 | 11 | 12 | 13 | 14 | 15 | 16 | 17 | 18 | 19 | 20 | 21 | 22 | 23 | 24 |
|---|---|---|---|---|---|---|---|---|---|---|---|---|---|---|---|---|---|---|---|---|---|---|---|
| 576 |  |  |  | 5 |  |  |  |  |  |  |  |  |  |  |  |  |  |  |  | 20 |  |  | 553 |
|  | 551 |  | 28 |  |  |  |  |  |  |  |  |  |  |  |  |  |  |  |  | 45 | 530 |  |  |
|  |  | 526 |  |  |  |  |  | 59 |  | 62 |  |  |  |  |  |  |  |  |  |  | 507 |  |  |
|  | 74 |  | 501 |  |  |  |  |  |  |  |  |  |  |  |  |  |  |  |  | 484 |  | 95 |  |
| 97 |  |  |  | 476 |  |  |  |  |  |  |  |  |  |  |  |  |  |  | 461 |  |  |  | 120 |
|  |  |  |  |  | 126 |  |  |  | 446 |  |  | 443 |  |  |  |  | 139 |  |  |  |  |  |  |
|  |  |  |  |  |  | 151 | 425 |  |  |  |  |  |  |  |  | 416 | 162 |  |  |  |  |  |  |
|  |  |  |  |  |  | 402 |  | 178 |  |  |  |  | 183 |  |  |  | 391 |  |  |  |  |  |  |
|  |  |  |  |  |  |  | 376 |  |  | 204 | 205 |  |  | 369 |  |  |  |  |  |  |  |  |  |
|  |  |  |  |  |  |  | 224 | 351 |  |  |  |  | 346 |  | 233 |  |  |  |  |  |  |  |  |
|  |  | 243 |  |  | 331 |  |  |  |  |  |  |  |  |  |  |  |  | 318 |  |  | 262 |  |  |
|  |  |  |  |  |  |  | 273 |  |  | 301 | 300 |  |  | 280 |  |  |  |  |  |  |  |  |  |
|  |  |  |  |  |  |  | 297 |  |  | 277 | 276 |  |  | 304 |  |  |  |  |  |  |  |  |  |
|  |  | 315 |  |  | 259 |  |  |  |  |  |  |  |  |  |  |  |  | 246 |  |  | 334 |  |  |
|  |  |  |  |  |  |  | 344 | 231 |  |  |  |  | 226 | 353 |  |  |  |  |  |  |  |  |  |
|  |  |  |  |  |  |  | 208 |  |  | 372 | 373 |  |  | 201 |  |  |  |  |  |  |  |  |  |
|  |  |  |  |  | 186 |  |  | 394 |  |  |  |  | 399 |  |  | 175 |  |  |  |  |  |  |  |
|  |  |  |  |  |  | 415 | 161 |  |  |  |  |  |  |  | 152 | 426 |  |  |  |  |  |  |  |
|  |  |  |  |  | 438 |  |  |  | 134 |  |  | 131 |  |  |  |  | 451 |  |  |  |  |  |  |
| 457 |  |  | 116 |  |  |  |  |  |  |  |  |  |  |  |  |  |  | 101 |  |  |  |  | 480 |
|  | 482 |  | 93 |  |  |  |  |  |  |  |  |  |  |  |  |  |  |  |  | 76 | 503 |  |  |
|  |  | 70 |  |  |  |  |  |  | 515 |  |  | 518 |  |  |  |  |  |  |  |  | 51 |  |  |
|  | 47 |  | 532 |  |  |  |  |  |  |  |  |  |  |  |  |  |  |  |  | 549 |  | 26 |  |
| 24 |  |  | 557 |  |  |  |  |  |  |  |  |  |  |  |  |  |  | 572 |  |  |  |  | 1 |

Magic Square – Number Fill-ins [C] – Mathematical Formula 24 X 24 + 1 X 12 = 6924

# Solved

# [C]

Mathematical Formulas

$8 \times 8 + 1 \times 4 = 260$

$12 \times 12 + 1 \times 6 = 870$

$16 \times 16 + 1 \times 8 = 2056$

$20 \times 20 + 1 \times 10 = 4010$

$24 \times 24 + 1 \times 12 = 6924$

Magic Square
Number Fill-ins [C]
Mathematical Formula

# SHELSEAS MAGIC SQUARE II
## It Works

Positive [C] Diagonals are shown for the same sum of 260

| 64 |    |    | 61 | 60 |    |    | 57 |
|----|----|----|----|----|----|----|----|
|    | 55 | 54 |    |    | 51 | 50 |    |
|    | 47 | 46 |    |    | 43 | 42 |    |
| 40 |    |    | 37 | 36 |    |    | 33 |
| 32 |    |    | 29 | 28 |    |    | 25 |
|    | 23 | 22 |    |    | 19 | 18 |    |
|    | 15 | 14 |    |    | 11 | 10 |    |
| 8  |    |    | 5  | 4  |    |    | 1  |

Negative [C]

|    | 2  | 3  |    |    | 6  | 7  |    |
|----|----|----|----|----|----|----|----|
| 9  |    |    | 12 | 13 |    |    | 16 |
| 17 |    |    | 20 | 21 |    |    | 24 |
|    | 26 | 27 |    |    | 30 | 31 |    |
|    | 34 | 35 |    |    | 38 | 39 |    |
| 41 |    |    | 44 | 45 |    |    | 48 |
| 49 |    |    | 52 | 53 |    |    | 56 |
|    | 58 | 59 |    |    | 62 | 63 |    |

Combination of both Positive & Negative number fill-ins [C] is shown for a solved Magic Square. Objective: Add each square of numbers in any row, column, or diagonal to proof the same sum of 260.

| 64 | 2  | 3  | 61 | 60 | 6  | 7  | 57 |
|----|----|----|----|----|----|----|----|
| 9  | 55 | 54 | 12 | 13 | 51 | 50 | 16 |
| 17 | 47 | 46 | 20 | 21 | 43 | 42 | 24 |
| 40 | 26 | 27 | 37 | 36 | 30 | 31 | 33 |
| 32 | 34· | 35 | 29 | 28 | 38 | 39 | 25 |
| 41 | 23 | 22 | 44 | 45 | 19 | 18 | 48 |
| 49 | 15 | 14 | 52 | 53 | 11 | 10 | 56 |
| 8  | 58 | 59 | 5  | 4  | 62 | 63 | 1  |

Magic Square
Number Fill-ins [C]
Mathematical Formula 8 X 8 + 1 X 4 = 260

# SHELSEAS MAGIC SQUARE II
## It Works

### Positive [C]

| 144 | 143 | 142 |     |     |     |     |     |     | 135 | 134 | 133 |
|-----|-----|-----|-----|-----|-----|-----|-----|-----|-----|-----|-----|
| 132 |     |     | 129 | 128 |     |     | 125 | 124 |     |     | 121 |
| 120 |     | 118 | 117 |     |     |     |     | 112 | 111 |     | 109 |
|     | 107 | 106 |     |     | 103 | 102 |     |     | 99  | 98  |     |
|     | 95  |     |     | 92  | 91  | 90  | 89  |     |     | 86  |     |
|     |     | 81  | 80  | 79  | 78  | 77  | 76  |     |     |     |     |
|     |     | 69  | 68  | 67  | 66  | 65  | 64  |     |     |     |     |
|     | 59  |     |     | 56  | 55  | 54  | 53  |     |     | 50  |     |
|     | 47  | 46  |     |     | 43  | 42  |     |     | 39  | 38  |     |
| 36  |     | 34  | 33  |     |     |     |     | 28  | 27  |     | 25  |
| 24  |     |     | 21  | 20  |     |     | 17  | 16  |     |     | 13  |
| 12  | 11  | 10  |     |     |     |     |     |     | 3   | 2   | 1   |

### Negative [C]

|     |     |     | 4   | 5   | 6   | 7   | 8   | 9   |     |     |     |
|-----|-----|-----|-----|-----|-----|-----|-----|-----|-----|-----|-----|
|     | 14  | 15  |     |     | 18  | 19  |     |     | 22  | 23  |     |
|     | 26  |     |     | 29  | 30  | 31  | 32  |     |     | 35  |     |
| 37  |     |     | 40  | 41  |     |     | 44  | 45  |     |     | 48  |
| 49  |     | 51  | 52  |     |     |     |     | 57  | 58  |     | 60  |
| 61  | 62  | 63  |     |     |     |     |     |     | 70  | 71  | 72  |
| 73  | 74  | 75  |     |     |     |     |     |     | 82  | 83  | 84  |
| 85  |     | 87  | 88  |     |     |     |     | 93  | 94  |     | 96  |
| 97  |     |     | 100 | 101 |     |     | 104 | 105 |     |     | 108 |
|     | 110 |     |     | 113 | 114 | 115 | 116 |     |     | 119 |     |
|     | 122 | 123 |     |     | 126 | 127 |     |     | 130 | 131 |     |
|     |     |     | 136 | 137 | 138 | 139 | 140 | 141 |     |     |     |

Combination of both Positive & Negative number fill-ins [C] is shown for a solved Magic Square. Objective: Add each square of numbers in any row, column or diagonal to proof the same sum of 870.

| 144 | 143 | 142 | 4   | 5   | 6   | 7   | 8   | 9   | 135 | 134 | 133 |
|-----|-----|-----|-----|-----|-----|-----|-----|-----|-----|-----|-----|
| 132 | 14  | 15  | 129 | 128 | 18  | 19  | 125 | 124 | 22  | 23  | 121 |
| 120 | 26  | 118 | 117 | 29  | 30  | 31  | 32  | 112 | 111 | 35  | 109 |
| 37  | 107 | 106 | 40  | 41  | 103 | 102 | 44  | 45  | 99  | 98  | 48  |
| 49  | 95  | 51  | 52  | 92  | 91  | 90  | 89  | 57  | 58  | 86  | 60  |
| 61  | 62  | 63  | 81  | 80  | 79  | 78  | 77  | 76  | 70  | 71  | 72  |
| 73  | 74  | 75  | 69  | 68  | 67  | 66  | 65  | 64  | 82  | 83  | 84  |
| 85  | 59  | 87  | 88  | 56  | 55  | 54  | 53  | 93  | 94  | 50  | 96  |
| 97  | 47  | 46  | 100 | 101 | 43  | 42  | 104 | 105 | 39  | 38  | 108 |
| 36  | 110 | 34  | 33  | 113 | 114 | 115 | 116 | 28  | 27  | 119 | 25  |
| 24  | 122 | 123 | 21  | 20  | 126 | 127 | 17  | 16  | 130 | 131 | 13  |
| 12  | 11  | 10  | 136 | 137 | 138 | 139 | 140 | 141 | 3   | 2   | 1   |

Magic Square – Number Fill-ins [C] – Mathematical Formula 12 X 12 + 1 X 6 = 870

# SHELSEAS MAGIC SQUARE II
## It Works

### Positive [C] Diagonals are shown for the same sum of 2056

| | | | | | | | | | | | | | | | |
|---|---|---|---|---|---|---|---|---|---|---|---|---|---|---|---|
| 256 | 255 | 254 | 253 | | | | | | | | | 244 | 243 | 242 | 241 |
| 240 | 239 | 238 | | | 235 | | | | | 230 | | | 227 | 226 | 225 |
| 224 | 223 | 222 | 221 | | | | | | | | | 212 | 211 | 210 | 209 |
| 208 | | 206 | 205 | | | 202 | | | 199 | | | 196 | 195 | | 193 |
| | | | | 188 | 187 | 186 | 185 | 184 | 183 | 182 | 181 | | | | |
| | 175 | | | 172 | 171 | | 169 | 168 | | 166 | 165 | | | 162 | |
| | | | 157 | 156 | | 154 | 153 | 152 | 151 | | 149 | 148 | | | |
| | | | | 140 | 139 | 138 | 137 | 136 | 135 | 134 | 133 | | | | |
| | | | | 124 | 123 | 122 | 121 | 120 | 119 | 118 | 117 | | | | |
| | | | 109 | 108 | | 106 | 105 | 104 | 103 | | 101 | 100 | | | |
| | 95 | | | 92 | 91 | | 89 | 88 | | 86 | 85 | | | 82 | |
| | | | | 76 | 75 | 74 | 73 | 72 | 71 | 70 | 69 | | | | |
| 64 | | 62 | 61 | | | 58 | | | 55 | | | 52 | 51 | | 49 |
| 48 | 47 | 46 | 45 | | | | | | | | | 36 | 35 | 34 | 33 |
| 32 | 31 | 30 | | | 27 | | | | | 22 | | | 19 | 18 | 17 |
| 16 | 15 | 14 | 13 | | | | | | | | | 4 | 3 | 2 | 1 |

### Negative [C]

| | | | | | | | | | | | | | | | |
|---|---|---|---|---|---|---|---|---|---|---|---|---|---|---|---|
| | | | | 5 | 6 | 7 | 8 | 9 | 10 | 11 | 12 | | | | |
| | | | 20 | 21 | | 23 | 24 | 25 | 26 | | 28 | 29 | | | |
| | | | | 37 | 38 | 39 | 40 | 41 | 42 | 43 | 44 | | | | |
| | 50 | | | 53 | 54 | | 56 | 57 | | 59 | 60 | | | 63 | |
| 65 | 66 | 67 | 68 | | | | | | | | | 77 | 78 | 79 | 80 |
| 81 | | 83 | 84 | | | 87 | | | 90 | | | 93 | 94 | | 96 |
| 97 | 98 | 99 | | | 102 | | | | | 107 | | | 110 | 111 | 112 |
| 113 | 114 | 115 | 116 | | | | | | | | | 125 | 126 | 127 | 128 |
| 129 | 130 | 131 | 132 | | | | | | | | | 141 | 142 | 143 | 144 |
| 145 | 146 | 147 | | | 150 | | | | | 155 | | | 158 | 159 | 160 |
| 161 | | 163 | 164 | | | 167 | | | 170 | | | 173 | 174 | | 176 |
| 177 | 178 | 179 | 180 | | | | | | | | | 189 | 190 | 191 | 192 |
| | 194 | | | 197 | 198 | | 200 | 201 | | 203 | 204 | | | 207 | |
| | | | | 213 | 214 | 215 | 216 | 217 | 218 | 219 | 220 | | | | |
| | | | 228 | 229 | | 231 | 232 | 233 | 234 | | 236 | 237 | | | |
| | | | | 245 | 246 | 247 | 248 | 249 | 250 | 251 | 252 | | | | |

### Magic Square
### Number Fill-ins [C]
Mathematical Formula 16 X 16 + 1 X 8 = 2056

# SHELSEAS MAGIC SQUARE II
## It Works

Combination of both Positive & Negative number fill-ins [C] is shown for a solved
Magic Square.  Objective:  Add each square of numbers in any row, column, or diagonal
to proof the same sum of 2056.

| 256 | 255 | 254 | 253 | 5 | 6 | 7 | 8 | 9 | 10 | 11 | 12 | 244 | 243 | 242 | 241 |
|-----|-----|-----|-----|-----|-----|-----|-----|-----|-----|-----|-----|-----|-----|-----|-----|
| 240 | 239 | 238 | 20 | 21 | 235 | 23 | 24 | 25 | 26 | 230 | 28 | 29 | 227 | 226 | 225 |
| 224 | 223 | 222 | 221 | 37 | 38 | 39 | 40 | 41 | 42 | 43 | 44 | 212 | 211 | 210 | 209 |
| 208 | 50 | 206 | 205 | 53 | 54 | 202 | 56 | 57 | 199 | 59 | 60 | 196 | 195 | 63 | 193 |
| 65 | 66 | 67 | 68 | 188 | 187 | 186 | 185 | 184 | 183 | 182 | 181 | 77 | 78 | 79 | 80 |
| 81 | 175 | 83 | 84 | 172 | 171 | 87 | 169 | 168 | 90 | 166 | 165 | 93 | 94 | 162 | 96 |
| 97 | 98 | 99 | 157 | 156 | 102 | 154 | 153 | 152 | 151 | 107 | 149 | 148 | 110 | 111 | 112 |
| 113 | 114 | 115 | 116 | 140 | 139 | 138 | 137 | 136 | 135 | 134 | 133 | 125 | 126 | 127 | 128 |
| 129 | 130 | 131 | 132 | 124 | 123 | 122 | 121 | 120 | 119 | 118 | 117 | 141 | 142 | 143 | 144 |
| 145 | 146 | 147 | 109 | 108 | 150 | 106 | 105 | 104 | 103 | 155 | 101 | 100 | 158 | 159 | 160 |
| 161 | 95 | 163 | 164 | 92 | 91 | 167 | 89 | 88 | 170 | 86 | 85 | 173 | 174 | 82 | 176 |
| 177 | 178 | 179 | 180 | 76 | 75 | 74 | 73 | 72 | 71 | 70 | 69 | 189 | 190 | 191 | 192 |
| 64 | 194 | 62 | 61 | 197 | 198 | 58 | 200 | 201 | 55 | 203 | 204 | 52 | 51 | 207 | 49 |
| 48 | 47 | 46 | 45 | 213 | 214 | 215 | 216 | 217 | 218 | 219 | 220 | 36 | 35 | 34 | 33 |
| 32 | 31 | 30 | 228 | 229 | 27 | 231 | 232 | 233 | 234 | 22 | 236 | 237 | 19 | 18 | 17 |
| 16 | 15 | 14 | 13 | 245 | 246 | 247 | 248 | 249 | 250 | 251 | 252 | 4 | 3 | 2 | 1 |

# 2056

Magic Square
Number Fill-ins [C]
Mathematical Formula 16 X 16 + 1 X 8 = 2056

# SHELSEAS MAGIC SQUARE II
## It Works

### Positive [C]

| | | | | | | | | | | | | | | | | | | | |
|---|---|---|---|---|---|---|---|---|---|---|---|---|---|---|---|---|---|---|---|
| 400 | 399 | 398 | 397 | 396 | | | | | | | | | | | 385 | 384 | 383 | 382 | 381 |
| 380 | 379 | 378 | 377 | | | | | 372 | | | 369 | | | | | 364 | 363 | 362 | 361 |
| 360 | 359 | 358 | 357 | | | | | 352 | | | 349 | | | | | 344 | 343 | 342 | 341 |
| 340 | 339 | 338 | | | | | 333 | 332 | | | 329 | 328 | | | | | 323 | 322 | 321 |
| 320 | | | | 316 | 315 | 314 | 313 | | | | | 308 | 307 | 306 | 305 | | | | 301 |
| | | | | 296 | 295 | 294 | 293 | | 291 | 290 | | 288 | 287 | 286 | 285 | | | | |
| | | | | 276 | 275 | 274 | 273 | | 271 | 270 | | 268 | 267 | 266 | 265 | | | | |
| | | | 257 | 256 | 255 | 254 | | | 251 | 250 | | | 247 | 246 | 245 | 244 | | | |
| | 239 | 238 | 237 | | | | | 232 | 231 | 230 | 229 | | | | | 224 | 223 | 222 | |
| | | | | | 215 | 214 | 213 | 212 | 211 | 210 | 209 | 208 | 207 | 206 | | | | | |
| | | | | | 195 | 194 | 193 | 192 | 191 | 190 | 189 | 188 | 187 | 186 | | | | | |
| | 179 | 178 | 177 | | | | | 172 | 171 | 170 | 169 | | | | | 164 | 163 | 162 | |
| | | | 157 | 156 | 155 | 154 | | | 151 | 150 | | | 147 | 146 | 145 | 144 | | | |
| | | | | 136 | 135 | 134 | 133 | | 131 | 130 | | 128 | 127 | 126 | 125 | | | | |
| | | | | 116 | 115 | 114 | 113 | | 111 | 110 | | 108 | 107 | 106 | 105 | | | | |
| 100 | | | | 96 | 95 | 94 | 93 | | | | | 88 | 87 | 86 | 85 | | | | 81 |
| 80 | 79 | 78 | | | | | 73 | 72 | | | 69 | 68 | | | | | 63 | 62 | 61 |
| 60 | 59 | 58 | 57 | | | | | 52 | | | 49 | | | | | 44 | 43 | 42 | 41 |
| 40 | 39 | 38 | 37 | | | | | 32 | | | 29 | | | | | 24 | 23 | 22 | 21 |
| 20 | 19 | 18 | 17 | 16 | | | | | | | | | | | 5 | 4 | 3 | 2 | 1 |

### Negative [C]

| | | | | | | | | | | | | | | | | | | | |
|---|---|---|---|---|---|---|---|---|---|---|---|---|---|---|---|---|---|---|---|
| | | | | | 6 | 7 | 8 | 9 | 10 | 11 | 12 | 13 | 14 | 15 | | | | | |
| | | | | 25 | 26 | 27 | 28 | | 30 | 31 | | 33 | 34 | 35 | 36 | | | | |
| | | | | 45 | 46 | 47 | 48 | | 50 | 51 | | 53 | 54 | 55 | 56 | | | | |
| | | | 64 | 65 | 66 | 67 | | | 70 | 71 | | | 74 | 75 | 76 | 77 | | | |
| | 82 | 83 | 84 | | | | | 89 | 90 | 91 | 92 | | | | | 97 | 98 | 99 | |
| 101 | 102 | 103 | 104 | | | | | 109 | | | 112 | | | | | 117 | 118 | 119 | 120 |
| 121 | 122 | 123 | 124 | | | | | 129 | | | 132 | | | | | 137 | 138 | 139 | 140 |
| 141 | 142 | 143 | | | | | 148 | 149 | | | 152 | 153 | | | | | 158 | 159 | 160 |
| 161 | | | | 165 | 166 | 167 | 168 | | | | | 173 | 174 | 175 | 176 | | | | 180 |
| 181 | 182 | 183 | 184 | 185 | | | | | | | | | | | 196 | 197 | 198 | 199 | 200 |
| 201 | 202 | 203 | 204 | 205 | | | | | | | | | | | 216 | 217 | 218 | 219 | 220 |
| 221 | | | | 225 | 226 | 227 | 228 | | | | | 233 | 234 | 235 | 236 | | | | 240 |
| 241 | 242 | 243 | | | | | 248 | 249 | | | 252 | 253 | | | | | 258 | 259 | 260 |
| 261 | 262 | 263 | 264 | | | | | 269 | | | 272 | | | | | 277 | 278 | 279 | 280 |
| 281 | 282 | 283 | 284 | | | | | 289 | | | 292 | | | | | 297 | 298 | 299 | 300 |
| | 302 | 303 | 304 | | | | | 309 | 310 | 311 | 312 | | | | | 317 | 318 | 319 | |
| | | | 324 | 325 | 326 | 327 | | | 330 | 331 | | | 334 | 335 | 336 | 337 | | | |
| | | | | 345 | 346 | 347 | 348 | | 350 | 351 | | 353 | 354 | 355 | 356 | | | | |
| | | | | 365 | 366 | 367 | 368 | | 370 | 371 | | 373 | 374 | 375 | 376 | | | | |
| | | | | | 386 | 387 | 388 | 389 | 390 | 391 | 392 | 393 | 394 | 395 | | | | | |

Magic Square – Number Fill-ins [C] – Mathematical Formula 20 X 20 + 1 X 10 = 4010

# SHELSEAS MAGIC SQUARE II
## It Works

Combination of both Positive & Negative number fill-ins [C] is shown for a solved Magic Square.  Objective:  Add each square of numbers in any row, column, or diagonal to proof the same sum of 4010.

| | | | | | | | | | | | | | | | | | | | |
|---|---|---|---|---|---|---|---|---|---|---|---|---|---|---|---|---|---|---|---|
| 400 | 399 | 398 | 397 | 396 | 6 | 7 | 8 | 9 | 10 | 11 | 12 | 13 | 14 | 15 | 385 | 384 | 383 | 382 | 381 |
| 380 | 379 | 378 | 377 | 25 | 26 | 27 | 28 | 372 | 30 | 31 | 369 | 33 | 34 | 35 | 36 | 364 | 363 | 362 | 361 |
| 360 | 359 | 358 | 357 | 45 | 46 | 47 | 48 | 352 | 50 | 51 | 349 | 53 | 54 | 55 | 56 | 344 | 343 | 342 | 341 |
| 340 | 339 | 338 | 64 | 65 | 66 | 67 | 333 | 332 | 70 | 71 | 329 | 328 | 74 | 75 | 76 | 77 | 323 | 322 | 321 |
| 320 | 82 | 83 | 84 | 316 | 315 | 314 | 313 | 89 | 90 | 91 | 92 | 308 | 307 | 306 | 305 | 97 | 98 | 99 | 301 |
| 101 | 102 | 103 | 104 | 296 | 295 | 294 | 293 | 109 | 291 | 290 | 112 | 288 | 287 | 286 | 285 | 117 | 118 | 119 | 120 |
| 121 | 122 | 123 | 124 | 276 | 275 | 274 | 273 | 129 | 271 | 270 | 132 | 268 | 267 | 266 | 265 | 137 | 138 | 139 | 140 |
| 141 | 142 | 143 | 257 | 256 | 255 | 254 | 148 | 149 | 251 | 250 | 152 | 153 | 247 | 246 | 245 | 244 | 158 | 159 | 160 |
| 161 | 239 | 238 | 237 | 165 | 166 | 167 | 168 | 232 | 231 | 230 | 229 | 173 | 174 | 175 | 176 | 224 | 223 | 222 | 180 |
| 181 | 182 | 183 | 184 | 185 | 215 | 214 | 213 | 212 | 211 | 210 | 209 | 208 | 207 | 206 | 196 | 197 | 198 | 199 | 200 |
| 201 | 202 | 203 | 204 | 205 | 195 | 194 | 193 | 192 | 191 | 190 | 189 | 188 | 187 | 186 | 216 | 217 | 218 | 219 | 220 |
| 221 | 179 | 178 | 177 | 225 | 226 | 227 | 228 | 172 | 171 | 170 | 169 | 233 | 234 | 235 | 236 | 164 | 163 | 162 | 240 |
| 241 | 242 | 243 | 157 | 156 | 155 | 154 | 248 | 249 | 151 | 150 | 252 | 253 | 147 | 146 | 145 | 144 | 258 | 259 | 260 |
| 261 | 262 | 263 | 264 | 136 | 135 | 134 | 133 | 269 | 131 | 130 | 272 | 128 | 127 | 126 | 125 | 277 | 278 | 279 | 280 |
| 281 | 282 | 283 | 284 | 116 | 115 | 114 | 113 | 289 | 111 | 110 | 292 | 108 | 107 | 106 | 105 | 297 | 298 | 299 | 300 |
| 100 | 302 | 303 | 304 | 96 | 95 | 94 | 93 | 309 | 310 | 311 | 312 | 88 | 87 | 86 | 85 | 317 | 318 | 319 | 81 |
| 80 | 79 | 78 | 324 | 325 | 326 | 327 | 73 | 72 | 330 | 331 | 69 | 68 | 334 | 335 | 336 | 337 | 63 | 62 | 61 |
| 60 | 59 | 58 | 57 | 345 | 346 | 347 | 348 | 52 | 350 | 351 | 49 | 353 | 354 | 355 | 356 | 44 | 43 | 42 | 41 |
| 40 | 39 | 38 | 37 | 365 | 366 | 367 | 368 | 32 | 370 | 371 | 29 | 373 | 374 | 375 | 376 | 24 | 23 | 22 | 21 |
| 20 | 19 | 18 | 17 | 16 | 386 | 387 | 388 | 389 | 390 | 391 | 392 | 393 | 394 | 395 | 5 | 4 | 3 | 2 | 1 |

# 4010

Magic Square
Number Fill-ins [C]
Mathematical Formula 20 X 20 + 1 X 10 = 4010

# SHELSEAS MAGIC SQUARE II
## It Works

### Positive [C]

|  |  |  |  |  |  |  |  |  |  |  |  |  |  |  |  |  |  |  |  |  |  |  |  |
|---|---|---|---|---|---|---|---|---|---|---|---|---|---|---|---|---|---|---|---|---|---|---|---|
| 576 | 575 | 574 | 573 |  |  |  |  |  |  | 566 | 565 | 564 | 563 |  |  |  |  |  |  | 556 | 555 | 554 | 553 |
| 552 | 551 | 550 |  |  | 547 | 546 |  |  |  |  | 541 | 540 |  |  |  |  | 535 | 534 |  |  | 531 | 530 | 529 |
| 528 | 527 | 526 | 525 | 524 | 523 |  |  |  |  |  |  |  |  |  |  |  |  | 510 | 509 | 508 | 507 | 506 | 505 |
| 504 |  | 502 | 501 | 500 |  |  | 497 | 496 |  |  |  |  |  |  | 489 | 488 |  |  | 485 | 484 | 483 |  | 481 |
|  |  | 478 | 477 | 476 | 475 | 474 | 473 |  |  |  |  |  |  |  |  | 464 | 463 | 462 | 461 | 460 | 459 |  |  |
|  | 455 | 454 |  | 452 |  | 450 |  |  | 447 | 446 |  |  | 443 | 442 |  |  | 439 |  | 437 |  | 435 | 434 |  |
|  | 431 |  |  | 428 | 427 |  | 425 | 424 | 423 |  |  |  |  | 418 | 417 | 416 |  | 414 | 413 |  |  | 410 |  |
|  |  |  | 405 | 404 |  | 402 | 401 | 400 |  |  | 397 | 396 |  |  | 393 | 392 | 391 |  | 389 | 388 |  |  |  |
|  |  |  | 381 |  |  | 378 | 377 | 376 | 375 | 374 |  |  | 371 | 370 | 369 | 368 | 367 |  |  | 364 |  |  |  |
|  |  |  |  |  | 355 | 354 |  | 352 | 351 | 350 | 349 | 348 | 347 | 346 | 345 |  | 343 | 342 |  |  |  |  |  |
| 336 |  |  |  |  | 331 |  |  | 328 | 327 | 326 | 325 | 324 | 323 | 322 | 321 |  |  | 318 |  |  |  |  | 313 |
| 312 | 311 |  |  |  |  |  | 305 |  | 303 | 302 | 301 | 300 | 299 | 298 |  | 296 |  |  |  |  |  | 290 | 289 |
| 288 | 287 |  |  |  |  |  | 281 |  | 279 | 278 | 277 | 276 | 275 | 274 |  | 272 |  |  |  |  |  | 266 | 265 |
| 264 |  |  |  |  | 259 |  |  | 256 | 255 | 254 | 253 | 252 | 251 | 250 | 249 |  |  | 246 |  |  |  |  | 241 |
|  |  |  |  |  | 235 | 234 |  | 232 | 231 | 230 | 229 | 228 | 227 | 226 | 225 |  | 223 | 222 |  |  |  |  |  |
|  |  |  | 213 |  |  | 210 | 209 | 208 | 207 | 206 |  |  | 203 | 202 | 201 | 200 | 199 |  |  | 196 |  |  |  |
|  |  |  | 189 | 188 |  | 186 | 185 | 184 |  |  | 181 | 180 |  |  | 177 | 176 | 175 |  | 173 | 172 |  |  |  |
|  | 167 |  |  | 164 | 163 |  | 161 | 160 | 159 |  |  |  |  | 154 | 153 | 152 |  | 150 | 149 |  |  | 146 |  |
|  | 143 | 142 |  | 140 |  | 138 |  |  | 135 | 134 |  |  | 131 | 130 |  |  | 127 |  | 125 |  | 123 | 122 |  |
|  |  | 118 | 117 | 116 | 115 | 114 | 113 |  |  |  |  |  |  |  |  | 104 | 103 | 102 | 101 | 100 | 99 |  |  |
| 96 |  | 94 | 93 | 92 |  |  | 89 | 88 |  |  |  |  |  |  | 81 | 80 |  |  | 77 | 76 | 75 |  | 73 |
| 72 | 71 | 70 | 69 | 68 | 67 |  |  |  |  |  |  |  |  |  |  |  |  | 54 | 53 | 52 | 51 | 50 | 49 |
| 48 | 47 | 46 |  |  | 43 | 42 |  |  |  |  | 37 | 36 |  |  |  |  | 31 | 30 |  |  | 27 | 26 | 25 |
| 24 | 23 | 22 | 21 |  |  |  |  |  |  | 14 | 13 | 12 | 11 |  |  |  |  |  |  | 4 | 3 | 2 | 1 |

### Negative [C]

|  |  |  |  |  |  |  |  |  |  |  |  |  |  |  |  |  |  |  |  |  |  |  |  |
|---|---|---|---|---|---|---|---|---|---|---|---|---|---|---|---|---|---|---|---|---|---|---|---|
|  |  |  |  | 5 | 6 | 7 | 8 | 9 | 10 |  |  |  |  | 15 | 16 | 17 | 18 | 19 | 20 |  |  |  |  |
|  |  |  | 28 | 29 |  |  | 32 | 33 | 34 | 35 |  |  | 38 | 39 | 40 | 41 |  |  | 44 | 45 |  |  |  |
|  |  |  |  |  |  | 55 | 56 | 57 | 58 | 59 | 60 | 61 | 62 | 63 | 64 | 65 | 66 |  |  |  |  |  |  |
|  | 74 |  |  |  | 78 | 79 |  |  | 82 | 83 | 84 | 85 | 86 | 87 |  |  | 90 | 91 |  |  |  | 95 |  |
| 97 | 98 |  |  |  |  |  |  | 105 | 106 | 107 | 108 | 109 | 110 | 111 | 112 |  |  |  |  |  |  | 119 | 120 |
| 121 |  |  | 124 |  | 126 |  | 128 | 129 |  |  | 132 | 133 |  |  | 136 | 137 |  | 139 |  | 141 |  |  | 144 |
| 145 |  | 147 | 148 |  |  | 151 |  |  |  | 155 | 156 | 157 | 158 |  |  |  | 162 |  |  | 165 | 166 |  | 168 |
| 169 | 170 | 171 |  |  | 174 |  |  |  | 178 | 179 |  |  | 182 | 183 |  |  |  | 187 |  |  | 190 | 191 | 192 |
| 193 | 194 | 195 |  | 197 | 198 |  |  |  |  |  | 204 | 205 |  |  |  |  |  | 211 | 212 |  | 214 | 215 | 216 |
| 217 | 218 | 219 | 220 | 221 |  |  | 224 |  |  |  |  |  |  |  |  | 233 |  |  | 236 | 237 | 238 | 239 | 240 |
|  | 242 | 243 | 244 | 245 |  | 247 | 248 |  |  |  |  |  |  |  |  | 257 | 258 |  | 260 | 261 | 262 | 263 |  |
|  |  | 267 | 268 | 269 | 270 | 271 |  | 273 |  |  |  |  |  |  | 280 |  | 282 | 283 | 284 | 285 | 286 |  |  |
|  |  | 291 | 292 | 293 | 294 | 295 |  | 297 |  |  |  |  |  |  | 304 |  | 306 | 307 | 308 | 309 | 310 |  |  |
|  | 314 | 315 | 316 | 317 |  | 319 | 320 |  |  |  |  |  |  |  |  | 329 | 330 |  | 332 | 333 | 334 | 335 |  |
| 337 | 338 | 339 | 340 | 341 |  |  | 344 |  |  |  |  |  |  |  |  | 353 |  |  | 356 | 357 | 358 | 359 | 360 |
| 361 | 362 | 363 |  | 365 | 366 |  |  |  |  |  | 372 | 373 |  |  |  |  |  | 379 | 380 |  | 382 | 383 | 384 |
| 385 | 386 | 387 |  |  | 390 |  |  |  | 394 | 395 |  |  | 398 | 399 |  |  |  | 403 |  |  | 406 | 407 | 408 |
| 409 |  | 411 | 412 |  |  | 415 |  |  |  | 419 | 420 | 421 | 422 |  |  |  | 426 |  |  | 429 | 430 |  | 432 |
| 433 |  |  | 436 |  | 438 |  | 440 | 441 |  |  | 444 | 445 |  |  | 448 | 449 |  | 451 |  | 453 |  |  | 456 |
| 457 | 458 |  |  |  |  |  |  | 465 | 466 | 467 | 468 | 469 | 470 | 471 | 472 |  |  |  |  |  |  | 479 | 480 |
|  | 482 |  |  |  | 486 | 487 |  |  | 490 | 491 | 492 | 493 | 494 | 495 |  |  | 498 | 499 |  |  |  | 503 |  |
|  |  |  |  |  |  | 511 | 512 | 513 | 514 | 515 | 516 | 517 | 518 | 519 | 520 | 521 | 522 |  |  |  |  |  |  |
|  |  |  | 532 | 533 |  |  | 536 | 537 | 538 | 539 |  |  | 542 | 543 | 544 | 545 |  |  | 548 | 549 |  |  |  |
|  |  |  |  | 557 | 558 | 559 | 560 | 561 | 562 |  |  |  |  | 567 | 568 | 569 | 570 | 571 | 572 |  |  |  |  |

Magic Square – Number Fill-ins [C] – Mathematical Formula 24 X 24 + 1 X 12 = 6924

# SHELSEAS MAGIC SQUARE II
## It Works

Combination of both Positive & Negative number fill-ins [C] is shown for a solved Magic Square. Objective: Add each square of numbers in any row, column, or diagonal to proof the same sum of 6924.

| | | | | | | | | | | | | | | | | | | | | | | | |
|---|---|---|---|---|---|---|---|---|---|---|---|---|---|---|---|---|---|---|---|---|---|---|---|
| 576 | 575 | 574 | 573 | 5 | 6 | 7 | 8 | 9 | 10 | 566 | 565 | 564 | 563 | 15 | 16 | 17 | 18 | 19 | 20 | 556 | 555 | 554 | 553 |
| 552 | 551 | 550 | 28 | 29 | 547 | 546 | 32 | 33 | 34 | 35 | 541 | 540 | 38 | 39 | 40 | 41 | 535 | 534 | 44 | 45 | 531 | 530 | 529 |
| 528 | 527 | 526 | 525 | 524 | 523 | 55 | 56 | 57 | 58 | 59 | 60 | 61 | 62 | 63 | 64 | 65 | 66 | 510 | 509 | 508 | 507 | 506 | 505 |
| 504 | 74 | 502 | 501 | 500 | 78 | 79 | 497 | 496 | 82 | 83 | 84 | 85 | 86 | 87 | 489 | 488 | 90 | 91 | 485 | 484 | 483 | 95 | 481 |
| 97 | 98 | 478 | 477 | 476 | 475 | 474 | 473 | 105 | 106 | 107 | 108 | 109 | 110 | 111 | 112 | 464 | 463 | 462 | 461 | 460 | 459 | 119 | 120 |
| 121 | 455 | 454 | 124 | 452 | 126 | 450 | 128 | 129 | 447 | 446 | 132 | 133 | 443 | 442 | 136 | 137 | 439 | 139 | 437 | 141 | 435 | 434 | 144 |
| 145 | 431 | 147 | 148 | 428 | 427 | 151 | 425 | 424 | 423 | 155 | 156 | 157 | 158 | 418 | 417 | 416 | 162 | 414 | 413 | 165 | 166 | 410 | 168 |
| 169 | 170 | 171 | 405 | 404 | 174 | 402 | 401 | 400 | 178 | 179 | 397 | 396 | 182 | 183 | 393 | 392 | 391 | 187 | 389 | 388 | 190 | 191 | 192 |
| 193 | 194 | 195 | 381 | 197 | 198 | 378 | 377 | 376 | 375 | 374 | 204 | 205 | 371 | 370 | 369 | 368 | 367 | 211 | 212 | 364 | 214 | 215 | 216 |
| 217 | 218 | 219 | 220 | 221 | 355 | 354 | 224 | 352 | 351 | 350 | 349 | 348 | 347 | 346 | 345 | 233 | 343 | 342 | 236 | 237 | 238 | 239 | 240 |
| 336 | 242 | 243 | 244 | 245 | 331 | 247 | 248 | 328 | 327 | 326 | 325 | 324 | 323 | 322 | 321 | 257 | 258 | 318 | 260 | 261 | 262 | 263 | 313 |
| 312 | 311 | 267 | 268 | 269 | 270 | 271 | 305 | 273 | 303 | 302 | 301 | 300 | 299 | 298 | 280 | 296 | 282 | 283 | 284 | 285 | 286 | 290 | 289 |
| 288 | 287 | 291 | 292 | 293 | 294 | 295 | 281 | 297 | 279 | 278 | 277 | 276 | 275 | 274 | 304 | 272 | 306 | 307 | 308 | 309 | 310 | 266 | 265 |
| 264 | 314 | 315 | 316 | 317 | 259 | 319 | 320 | 256 | 255 | 254 | 253 | 252 | 251 | 250 | 249 | 329 | 330 | 246 | 332 | 333 | 334 | 335 | 241 |
| 337 | 338 | 339 | 340 | 341 | 235 | 234 | 344 | 232 | 231 | 230 | 229 | 228 | 227 | 226 | 225 | 353 | 223 | 222 | 356 | 357 | 358 | 359 | 360 |
| 361 | 362 | 363 | 213 | 365 | 366 | 210 | 209 | 208 | 207 | 206 | 372 | 373 | 203 | 202 | 201 | 200 | 199 | 379 | 380 | 196 | 382 | 383 | 384 |
| 385 | 386 | 387 | 189 | 188 | 390 | 186 | 185 | 184 | 394 | 395 | 181 | 180 | 398 | 399 | 177 | 176 | 175 | 403 | 173 | 172 | 406 | 407 | 408 |
| 409 | 167 | 411 | 412 | 164 | 163 | 415 | 161 | 160 | 159 | 419 | 420 | 421 | 422 | 154 | 153 | 152 | 426 | 150 | 149 | 429 | 430 | 146 | 432 |
| 433 | 143 | 142 | 436 | 140 | 438 | 138 | 440 | 441 | 135 | 134 | 444 | 445 | 131 | 130 | 448 | 449 | 127 | 451 | 125 | 453 | 123 | 122 | 456 |
| 457 | 458 | 118 | 117 | 116 | 115 | 114 | 113 | 465 | 466 | 467 | 468 | 469 | 470 | 471 | 472 | 104 | 103 | 102 | 101 | 100 | 99 | 479 | 480 |
| 96 | 482 | 94 | 93 | 92 | 486 | 487 | 89 | 88 | 490 | 491 | 492 | 493 | 494 | 495 | 81 | 80 | 498 | 499 | 77 | 76 | 75 | 503 | 73 |
| 72 | 71 | 70 | 69 | 68 | 67 | 511 | 512 | 513 | 514 | 515 | 516 | 517 | 518 | 519 | 520 | 521 | 522 | 54 | 53 | 52 | 51 | 50 | 49 |
| 48 | 47 | 46 | 532 | 533 | 43 | 42 | 536 | 537 | 538 | 539 | 37 | 36 | 542 | 543 | 544 | 545 | 31 | 30 | 548 | 549 | 27 | 26 | 25 |
| 24 | 23 | 22 | 21 | 557 | 558 | 559 | 560 | 561 | 562 | 14 | 13 | 12 | 11 | 567 | 568 | 569 | 570 | 571 | 572 | 4 | 3 | 2 | 1 |

# 6924

Magic Square
Number Fill-ins [C]
Mathematical Formula 24 X 24 + 1 X 12 = 6924

The Number Fill-ins [D] Will Be Your Guideline
For A Solved Double Even – Magic Square.

# Number
# Fill-ins

# [D]

Positive [D] Start with the bottom row, left to right in each row, going up.  Begin with the number shown hint, count each square, only fill-in same shaded squares.

Negative [D] Start with the top row, right to left in each row, going down.  Begin with the number shown hint, count each square, only fill-in same shaded squares.

Objective:  Combine both Positive & Negative number fill-ins [D] in each Magic Square.  Add each square of numbers in any row, column, or diagonal to proof the same sum of integer for a solved Magic Square.

## Mathematical Formulas

$$8 \times 8 + 1 \times 4 = 260$$

$$12 \times 12 + 1 \times 6 = 870$$

$$16 \times 16 + 1 \times 8 = 2056$$

$$20 \times 20 + 1 \times 10 = 4010$$

$$24 \times 24 + 1 \times 12 = 6924$$

# SHELSEAS MAGIC SQUARE II
## It Works

Positive [D] Start with the bottom row, left to right in each row, going up. Begin with the number shown 1, count each square, only fill-in shaded squares.

| 13 |    |    | 16 |
|----|----|----|----|
|    | 10 | 11 |    |
|    | 6  | 7  |    |
| 1  |    |    | 4  |

Negative [D] Start with the top row, right to left in each row, going down. Begin with the number 2, count each square, only fill-in shaded squares.

|    | 3  | 2  |    |
|----|----|----|----|
| 8  |    |    | 5  |
| 12 |    |    | 9  |
|    | 15 | 14 |    |

Combination of both Positive & Negative number fill-ins [D] is shown for a solved Magic Square. Objective: Add each square of numbers in any row, column, or diagonal to proof the same sum of 34.

| 13 | 3  | 2  | 16 |
|----|----|----|----|
| 8  | 10 | 11 | 5  |
| 12 | 6  | 7  | 9  |
| 1  | 15 | 14 | 4  |

Magic Square
Number Fill-ins [D]
Mathematical Formula 4 X 4 + 1 X 2 = 34

# SHELSEAS MAGIC SQUARE II
## It Works

Positive [D] Start with the bottom row, left to right in each row, going up. Begin with the number 2, count each square, only fill-in shaded squares. Helpful hints below.

| | 58 | | | | | 63 | |
|---|---|---|---|---|---|---|---|
| 49 | | | | | | | 56 |
| | | | 44 | 45 | | | |
| | | 35 | | | 38 | | |
| | | 27 | | | 30 | | |
| | | | 20 | 21 | | | |
| 9 | | | | | | | 16 |
| | 2 | | | | | 7 | |

Negative [D) Start with the top row, right to left in each row, going down. Begin with the number 1, count each square, only fill-in shaded squares. Helpful hints below.

| 8 | | | | | | | 1 |
|---|---|---|---|---|---|---|---|
| | | | 13 | 12 | | | |
| | | 22 | | | 19 | | |
| | 31 | | | | | 26 | |
| | 39 | | | | | 34 | |
| | | 46 | | | 43 | | |
| | | | 53 | 52 | | | |
| 64 | | | | | | | 57 |

Objective: Combine both Positive & Negative number fill-ins [D]. Add each square of numbers in any row, column, or diagonal to proof the same sum of 260 for a solved Magic Square. Helpful hints below.

| 8 | 58 | | | | | 63 | 1 |
|---|---|---|---|---|---|---|---|
| 49 | | | 13 | 12 | | | 56 |
| | | 22 | 44 | 45 | 19 | | |
| | 31 | 35 | | | 38 | 26 | |
| | 39 | 27 | | | 30 | 34 | |
| | | 46 | 20 | 21 | 43 | | |
| 9 | | | 53 | 52 | | | 16 |
| 64 | 2 | | | | | 7 | 57 |

Magic Square – Number Fill-ins [D] – Mathematical Formula 8 X 8 + 1 X 4 = 260

# SHELSEAS MAGIC SQUARE II
## It Works

Positive [D] Start with the bottom row, left to right in each row, going up. Begin with the number 1, count each square, only fill-in shaded squares. Diagonals are shown for the same sum of 870. Helpful hints below.

| 1 | 2 | 3 | 4 | 5 | 6 | 7 | 8 | 9 | 10 | 11 | 12 |
|---|---|---|---|---|---|---|---|---|----|----|----|
| 133 |  |  |  |  |  |  |  |  |  |  | 144 |
|  | 122 |  |  |  |  |  |  |  |  | 131 |  |
|  |  | 111 |  |  |  |  |  |  | 118 |  |  |
|  |  |  | 100 |  |  |  |  | 105 |  |  |  |
|  |  |  |  | 89 |  | 92 |  |  |  |  |  |
|  |  |  |  | 78 | 79 |  |  |  |  |  |  |
|  |  |  |  | 66 | 67 |  |  |  |  |  |  |
|  |  |  |  | 53 |  | 56 |  |  |  |  |  |
|  |  |  | 40 |  |  |  |  | 45 |  |  |  |
|  |  | 27 |  |  |  |  |  |  | 34 |  |  |
|  | 14 |  |  |  |  |  |  |  |  | 23 |  |
| 1 |  |  |  |  |  |  |  |  |  |  | 12 |

Negative [D] Start with the top row, right to left in each row, going down. Begin with the number 4, count each square, only fill-in shaded squares. Helpful hints below.

| 1 | 2 | 3 | 4 | 5 | 6 | 7 | 8 | 9 | 10 | 11 | 12 |
|---|---|---|---|---|---|---|---|---|----|----|----|
|  |  |  | 9 |  |  |  |  | 4 |  |  |  |
|  |  |  |  | 20 |  |  | 17 |  |  |  |  |
|  |  |  |  | 31 | 30 |  |  |  |  |  |  |
| 48 |  |  |  |  |  |  |  |  |  |  | 37 |
|  | 59 |  |  |  |  |  |  |  |  | 50 |  |
|  |  | 70 |  |  |  |  |  |  | 63 |  |  |
|  |  | 82 |  |  |  |  |  |  | 75 |  |  |
|  | 95 |  |  |  |  |  |  |  |  | 86 |  |
| 108 |  |  |  |  |  |  |  |  |  |  | 97 |
|  |  |  |  | 115 | 114 |  |  |  |  |  |  |
|  |  |  | 128 |  |  |  | 125 |  |  |  |  |
|  |  | 141 |  |  |  |  |  | 136 |  |  |  |

Magic Square
Number Fill-ins [D]
Mathematical Formula 12 X 12 + 1 X 6 = 870

# SHELSEAS MAGIC SQUARE II
## It Works

Positive [D] Start with the bottom row, left to right in each row, going up. Begin with the number 1, count each square, only fill-in same shaded squares. Diagonals are shown for the same sum of 870.

Negative [D] Start with the top row, right to left in each row, going down. Begin with the number 4, count each square, only fill-in same shaded squares.

Objective: Combine both Positive & Negative number fill-ins [D]. Add each square of numbers in any row, column, or diagonal to proof the same sum of 870 for a solved Magic Square. Helpful hints below.

| | | | | | | | | | | | |
|-----|-----|-----|-----|-----|-----|-----|-----|-----|-----|-----|-----|
| 133 | | | 9 | | | | | 4 | | | 144 |
| | 122 | | | 20 | | | 17 | | | 131 | |
| | | 111 | | | 31 | 30 | | | 118 | | |
| 48 | | | 100 | | | | | 105 | | | 37 |
| | 59 | | | 89 | | | 92 | | | 50 | |
| | | 70 | | | 78 | 79 | | | 63 | | |
| | | 82 | | | 66 | 67 | | | 75 | | |
| | 95 | | | 53 | | | 56 | | | 86 | |
| 108 | | | 40 | | | | | 45 | | | 97 |
| | | 27 | | | 115 | 114 | | | 34 | | |
| | 14 | | | 128 | | | 125 | | | 23 | |
| 1 | | | 141 | | | | | 136 | | | 12 |

Magic Square
Number Fill-ins [D]
Mathematical Formula 12 X 12 + 1 X 6 = 870

# SHELSEAS MAGIC SQUARE II
## It Works

Positive [D] Start with the bottom row, left to right in each row, going up. Begin with the number 1, count each square, only fill-in shaded squares. Helpful hints below.

| | | | | | | | | | | | | | | | |
|---|---|---|---|---|---|---|---|---|---|---|---|---|---|---|---|
| 241 | | | | | | | | | | | | | | | 256 |
| | 227 | | | | | | | | | | | | 238 | | |
| | 210 | | | | | | | | | | | | | 223 | |
| | | | 197 | | | | | | | | 204 | | | | |
| | | 180 | | | | | | | | | | 189 | | | |
| | | | | | 167 | | | | 170 | | | | | | |
| | | | | 150 | | | | | | 155 | | | | | |
| | | | | | | 136 | 137 | | | | | | | | |
| | | | | | | 120 | 121 | | | | | | | | |
| | | | | 102 | | | | | | 107 | | | | | |
| | | | | | 87 | | | | 90 | | | | | | |
| | | | 68 | | | | | | | | 77 | | | | |
| | | 53 | | | | | | | | | | 60 | | | |
| | 34 | | | | | | | | | | | | | 47 | |
| | | 19 | | | | | | | | | | | 30 | | |
| 1 | | | | | | | | | | | | | | | 16 |

Negative [D] Start with the top row, right to left in each row, going down. Begin with the number 5, count each square, only fill-in shaded squares. Helpful hints below.

| | | | | | | | | | | | | | | | |
|---|---|---|---|---|---|---|---|---|---|---|---|---|---|---|---|
| | | | | 12 | | | | | | | 5 | | | | |
| | | | 29 | | | | | | | | | 20 | | | |
| | | | | | 42 | | | 39 | | | | | | | |
| | 63 | | | | | | | | | | | | | 50 | |
| 80 | | | | | | | | | | | | | | | 65 |
| | | | | | | 89 | 88 | | | | | | | | |
| | | 110 | | | | | | | | | | | 99 | | |
| | | | | 123 | | | | | | 118 | | | | | |
| | | | | 139 | | | | | | 134 | | | | | |
| | | 158 | | | | | | | | | | | 147 | | |
| | | | | | | 169 | 168 | | | | | | | | |
| 192 | | | | | | | | | | | | | | | 177 |
| | 207 | | | | | | | | | | | | | 194 | |
| | | | | | 218 | | | 215 | | | | | | | |
| | | | 237 | | | | | | | | | 228 | | | |
| | | | | 252 | | | | | | | 245 | | | | |

Magic Square – Number Fill-ins [D] – Mathematical Formula 16 X 16 + 1 X 8 = 2056

# SHELSEAS MAGIC SQUARE II
## It Works

Positive [D] Start with the bottom row, left to right in each row, going up. Begin with the number 1, count each square, only fill-in same shaded squares.

Negative [D] Start with the top row, right to left in each row, going down. Begin with the number 5, count each square, only fill-in same shaded squares.

Objective: Combine both Positive & Negative number fill-ins [D]. Add each square of numbers in any row, column, or diagonal to proof the same sum of 2056 for a solved Magic Square. Helpful hints below.

| | | | | | | | | | | | | | | | |
|---|---|---|---|---|---|---|---|---|---|---|---|---|---|---|---|
| 241 | | | | 12 | | | | | | | 5 | | | | 256 |
| | | 227 | 29 | | | | | | | | | 20 | 238 | | |
| | 210 | | | | | 42 | | | 39 | | | | | 223 | |
| | 63 | | | 197 | | | | | | | 204 | | | 50 | |
| 80 | | | 180 | | | | | | | | | 189 | | | 65 |
| | | | | | | 167 | 89 | 88 | 170 | | | | | | |
| | | 110 | | | 150 | | | | | 155 | | | 99 | | |
| | | | | | 123 | | 136 | 137 | | 118 | | | | | |
| | | | | | 139 | | 120 | 121 | | 134 | | | | | |
| | | 158 | | | 102 | | | | | 107 | | | 147 | | |
| | | | | | | 87 | 169 | 168 | 90 | | | | | | |
| 192 | | | 68 | | | | | | | | | 77 | | | 177 |
| | 207 | | | 53 | | | | | | | 60 | | | 194 | |
| | 34 | | | | | 218 | | | 215 | | | | | 47 | |
| | | 19 | 237 | | | | | | | | | 228 | 30 | | |
| 1 | | | | 252 | | | | | | | 245 | | | | 16 |

Magic Square
Number Fill-ins [D]
Mathematical Formula 16 X 16 + 1 X 8 = 2056

# SHELSEAS MAGIC SQUARE II
## It Works

Positive [D] Start with the bottom row, left to right in each row, going up. Begin with the number 1, count each square, only fill-in shaded squares. Helpful hints below.

| 1 | 2 | 3 | 4 | 5 | 6 | 7 | 8 | 9 | 10 | 11 | 12 | 13 | 14 | 15 | 16 | 17 | 18 | 19 | 20 |
|---|---|---|---|---|---|---|---|---|----|----|----|----|----|----|----|----|----|----|----|
| 381 |  |  |  |  |  |  |  |  |  |  |  |  |  |  |  |  |  |  | 400 |
|  | 362 |  |  |  |  |  |  |  |  |  |  |  |  |  |  |  |  | 379 |  |
|  |  |  | 345 |  |  |  |  |  |  |  |  |  |  |  | 356 |  |  |  |  |
|  |  |  |  |  | 327 |  |  |  |  |  |  |  | 334 |  |  |  |  |  |  |
|  |  | 303 |  |  |  |  |  |  |  |  |  |  |  |  |  | 318 |  |  |  |
|  |  |  |  |  |  | 288 |  |  |  |  | 293 |  |  |  |  |  |  |  |  |
|  |  |  |  | 264 |  |  |  |  |  |  |  |  |  |  | 277 |  |  |  |  |
|  |  |  |  |  | 246 |  |  |  |  |  |  |  | 255 |  |  |  |  |  |  |
|  |  |  |  |  |  |  | 229 |  |  | 232 |  |  |  |  |  |  |  |  |  |
|  |  |  |  |  |  |  |  | 210 | 211 |  |  |  |  |  |  |  |  |  |  |
|  |  |  |  |  |  |  |  | 190 | 191 |  |  |  |  |  |  |  |  |  |  |
|  |  |  |  |  |  |  | 169 |  |  | 172 |  |  |  |  |  |  |  |  |  |
|  |  |  |  |  | 146 |  |  |  |  |  |  |  | 155 |  |  |  |  |  |  |
|  |  |  |  | 124 |  |  |  |  |  |  |  |  |  |  | 137 |  |  |  |  |
|  |  |  |  |  |  | 108 |  |  |  |  | 113 |  |  |  |  |  |  |  |  |
|  |  | 83 |  |  |  |  |  |  |  |  |  |  |  |  |  | 98 |  |  |  |
|  |  |  |  |  | 67 |  |  |  |  |  |  | 74 |  |  |  |  |  |  |  |
|  |  |  | 45 |  |  |  |  |  |  |  |  |  |  | 56 |  |  |  |  |  |
|  | 22 |  |  |  |  |  |  |  |  |  |  |  |  |  |  |  |  | 39 |  |
| 1 |  |  |  |  |  |  |  |  |  |  |  |  |  |  |  |  |  |  | 20 |

Magic Square
Number Fill-ins [D]
Mathematical Formula 20 X 20 + 1 X 10 = 4010

# SHELSEAS MAGIC SQUARE II
## It Works

Negative [D] Start with the top row, right to left in each row, going down.  Begin with the number 5, count each square, only fill-in shaded squares.  Helpful hints below.

| 1 | 2 | 3 | 4 | 5 | 6 | 7 | 8 | 9 | 10 | 11 | 12 | 13 | 14 | 15 | 16 | 17 | 18 | 19 | 20 |
|---|---|---|---|---|---|---|---|---|----|----|----|----|----|----|----|----|----|----|----|
|  |  |  | 16 |  |  |  |  |  |  |  |  |  |  |  |  | 5 |  |  |  |
|  |  |  |  |  |  |  |  | 31 | 30 |  |  |  |  |  |  |  |  |  |  |
|  |  | 58 |  |  |  |  |  |  |  |  |  |  |  |  |  | 43 |  |  |  |
|  |  |  |  |  |  |  | 72 |  |  |  | 69 |  |  |  |  |  |  |  |  |
| 100 |  |  |  |  |  |  |  |  |  |  |  |  |  |  |  |  |  |  | 81 |
|  |  |  |  |  | 114 |  |  |  |  |  |  |  | 107 |  |  |  |  |  |  |
|  |  |  |  | 135 |  |  |  |  |  |  |  |  | 126 |  |  |  |  |  |  |
|  |  |  |  |  | 153 |  |  |  |  |  |  | 148 |  |  |  |  |  |  |  |
|  |  |  | 177 |  |  |  |  |  |  |  |  |  |  | 164 |  |  |  |  |  |
|  | 199 |  |  |  |  |  |  |  |  |  |  |  |  |  |  |  | 182 |  |  |
|  | 219 |  |  |  |  |  |  |  |  |  |  |  |  |  |  |  |  | 202 |  |
|  |  | 237 |  |  |  |  |  |  |  |  |  |  |  | 224 |  |  |  |  |  |
|  |  |  |  |  | 253 |  |  |  |  |  |  | 248 |  |  |  |  |  |  |  |
|  |  |  |  | 275 |  |  |  |  |  |  |  |  | 266 |  |  |  |  |  |  |
|  |  |  |  |  | 294 |  |  |  |  |  |  | 287 |  |  |  |  |  |  |  |
| 320 |  |  |  |  |  |  |  |  |  |  |  |  |  |  |  |  |  |  | 301 |
|  |  |  |  |  |  |  | 332 |  |  | 329 |  |  |  |  |  |  |  |  |  |
|  |  | 358 |  |  |  |  |  |  |  |  |  |  |  |  | 343 |  |  |  |  |
|  |  |  |  |  |  |  |  | 371 | 370 |  |  |  |  |  |  |  |  |  |  |
|  |  |  | 396 |  |  |  |  |  |  |  |  |  |  | 385 |  |  |  |  |  |

Magic Square
Number Fill-ins [D]
Mathematical Formula 20 X 20 + 1 X 10 = 4010

# SHELSEAS MAGIC SQUARE II
## It Works

Positive [D] Start with the bottom row, left to right in each row, going up.  Begin with the number 1, count each square, only fill-in same shaded squares.

Negative [D] Start with the top row, right to left in each row, going down.  Begin with the number 5, count each square, only fill-in same shaded squares.

Objective:  Combine both Positive & Negative number fill-ins [D].  Add each square of numbers in any row, column, or diagonal to proof the same sum of 4010 for a solved Magic Square.  Helpful hints below.

| | | | | | | | | | | | | | | | | | | | |
|---|---|---|---|---|---|---|---|---|---|---|---|---|---|---|---|---|---|---|---|
| 381 | | | 16 | | | | | | | | | | | | | | 5 | | 400 |
| | 362 | | | | | | 31 | 30 | | | | | | | | | | 379 | |
| | | 58 | | 345 | | | | | | | | | | | | 356 | | 43 | |
| | | | | | 327 | | 72 | | | 69 | 334 | | | | | | | | |
| 100 | | 303 | | | | | | | | | | | | | | | 318 | | 81 |
| | | | | | | 114 | 288 | | | | 293 | 107 | | | | | | | |
| | | | 264 | | 135 | | | | | | | | 126 | | 277 | | | | |
| | | | | | 246 | | 153 | | | | 148 | | 255 | | | | | | |
| | | | 177 | | | | | 229 | | 232 | | | | | 164 | | | | |
| | 199 | | | | | | | 210 | 211 | | | | | | | | 182 | | |
| | 219 | | | | | | | 190 | 191 | | | | | | | | 202 | | |
| | | | 237 | | | | | 169 | | | 172 | | | | 224 | | | | |
| | | | | | 146 | | 253 | | | | 248 | | 155 | | | | | | |
| | | | 124 | | 275 | | | | | | | | 266 | | 137 | | | | |
| | | | | | | 294 | 108 | | | | 113 | 287 | | | | | | | |
| 320 | | 83 | | | | | | | | | | | | | | | 98 | | 301 |
| | | | | | | 67 | 332 | | | 329 | | 74 | | | | | | | |
| | | 358 | | 45 | | | | | | | | | | | | 56 | 343 | | |
| | 22 | | | | | | | 371 | 370 | | | | | | | | | 39 | |
| 1 | | | | 396 | | | | | | | | | | | | 385 | | | 20 |

Magic Square – Number Fill-ins [D] – Mathematical Formula 20 X 20 + 1 X 10 = 4010

# SHELSEAS MAGIC SQUARE II
## It Works

Positive [D] Start with the bottom row, left to right in each row, going up. Begin with the number 1, count each square, only fill-in shaded squares. Diagonals are shown for the same sum of 6924. Helpful hints below.

| | | | | | | | | | | | | | | | | | | | | | | | |
|---|---|---|---|---|---|---|---|---|---|---|---|---|---|---|---|---|---|---|---|---|---|---|---|
| 553 | | | | | | | | | | | | | | | | | | | | | | | 576 |
| | 530 | | | | | | | | | | | | | | | | | | | | | 551 | |
| | | 507 | | | | | | | | | | | | | | | | | | | 526 | | |
| | | | 484 | | | | | | | | | | | | | | | | | 501 | | | |
| | | | | 461 | | | | | | | | | | | | | | | 476 | | | | |
| | | | | | 438 | | | | | | | | | | | | | 451 | | | | | |
| | | | | | | 415 | | | | | | | | | | | 426 | | | | | | |
| | | | | | | | 392 | | | | | | | | | 401 | | | | | | | |
| | | | | | | | | 369 | | | | | | | 376 | | | | | | | | |
| | | | | | | | | | 346 | | | | | 351 | | | | | | | | | |
| | | | | | | | | | | 323 | | | 326 | | | | | | | | | | |
| | | | | | | | | | | | 300 | 301 | | | | | | | | | | | |
| | | | | | | | | | | | 276 | 277 | | | | | | | | | | | |
| | | | | | | | | | | 251 | | | 254 | | | | | | | | | | |
| | | | | | | | | | 226 | | | | | 231 | | | | | | | | | |
| | | | | | | | | 201 | | | | | | | 208 | | | | | | | | |
| | | | | | | | 176 | | | | | | | | | 185 | | | | | | | |
| | | | | | | 151 | | | | | | | | | | | 162 | | | | | | |
| | | | | | 126 | | | | | | | | | | | | | 139 | | | | | |
| | | | | 101 | | | | | | | | | | | | | | | 116 | | | | |
| | | | 76 | | | | | | | | | | | | | | | | | 93 | | | |
| | | 51 | | | | | | | | | | | | | | | | | | | 70 | | |
| | 26 | | | | | | | | | | | | | | | | | | | | | 47 | |
| 1 | | | | | | | | | | | | | | | | | | | | | | | 24 |

Magic Square
Number Fill-ins [D]
Mathematical Formula 24 X 24 + 1 X 12 = 6924

# SHELSEAS MAGIC SQUARE II
## It Works

Negative [D] Start with the top row, right to left in each row, going down.  Begin with the number 6, count each square, only fill-in shaded squares.  Helpful hints below.

| | | | | | | | | | | | | | | | | | | | | | | | |
|---|---|---|---|---|---|---|---|---|---|---|---|---|---|---|---|---|---|---|---|---|---|---|---|
| | | | | | 19 | | | | | | | | | | | | | | 6 | | | | |
| | | | | 44 | | | | | | | | | | | | | | | | 29 | | | |
| | | | | | | | 65 | | | | | | | | | | 56 | | | | | | |
| | | | | | | | | | 87 | | | | | | 82 | | | | | | | | |
| | 119 | | | | | | | | | | | | | | | | | | | | | | 98 |
| 144 | | | | | | | | | | | | | | | | | | | | | | | 121 |
| | | | | | | | | | | 158 | | | | 155 | | | | | | | | | |
| | | 190 | | | | | | | | | | | | | | | | | | | | 171 | |
| | | | | | | | | | | | 205 | 204 | | | | | | | | | | | |
| | | | 237 | | | | | | | | | | | | | | | | | | 220 | | |
| | | | | | | 258 | | | | | | | | | | | | 247 | | | | | |
| | | | | | | | | 280 | | | | | | | | 273 | | | | | | | |
| | | | | | | | | 304 | | | | | | | | 297 | | | | | | | |
| | | | | | | 330 | | | | | | | | | | | | 319 | | | | | |
| | | | 357 | | | | | | | | | | | | | | | | | | 340 | | |
| | | | | | | | | | | | 373 | 372 | | | | | | | | | | | |
| | | 406 | | | | | | | | | | | | | | | | | | | | 387 | |
| | | | | | | | | | | 422 | | | | 419 | | | | | | | | | |
| 456 | | | | | | | | | | | | | | | | | | | | | | | 433 |
| | 479 | | | | | | | | | | | | | | | | | | | | | | 458 |
| | | | | | | | | | 495 | | | | | | 490 | | | | | | | | |
| | | | | | | | 521 | | | | | | | | | | 512 | | | | | | |
| | | | | 548 | | | | | | | | | | | | | | | | 533 | | | |
| | | | | | 571 | | | | | | | | | | | | | | 558 | | | | |

## Magic Square
## Number Fill-ins [D]
### Mathematical Formula 24 X 24 + 1 X 12 = 6924

# SHELSEAS MAGIC SQUARE II
## It Works

Positive [D] Start with the bottom row, left to right in each row, going up. Begin with the number 1, count each square, only fill-in same shaded squares.

Negative [D] Start with the top row, right to left in each row, going down. Begin with the number 6, count each square, only fill-in same shaded squares.

Objective:  Combine both Positive & Negative number fill-ins [D].  Add each square of numbers in any row, column, or diagonal to proof the same sum of 6924 for a solved Magic Square.  Helpful hints below.

| 1 | 2 | 3 | 4 | 5 | 6 | 7 | 8 | 9 | 10 | 11 | 12 | 13 | 14 | 15 | 16 | 17 | 18 | 19 | 20 | 21 | 22 | 23 | 24 |
|---|---|---|---|---|---|---|---|---|----|----|----|----|----|----|----|----|----|----|----|----|----|----|----|
| 553 | | | | | 19 | | | | | | | | | | | | | 6 | | | | | 576 |
| | 530 | | | 44 | | | | | | | | | | | | | | | 29 | | 551 | | |
| | | 507 | | | | | 65 | | | | | | | | | 56 | | | | | 526 | | |
| | | | 484 | | | | | | 87 | | | | | 82 | | | | | | | 501 | | |
| | 119 | | | 461 | | | | | | | | | | | | | | | 476 | | | 98 | |
| 144 | | | | | 438 | | | | | | | | | | | | | 451 | | | | | 121 |
| | | | | | | 415 | | | | 158 | | | 155 | | | | 426 | | | | | | |
| | 190 | | | | | 392 | | | | | | | | | | 401 | | | | | 171 | | |
| | | | | | | | 369 | | | | 205 | 204 | | | 376 | | | | | | | | |
| | | | 237 | | | | | | 346 | | | | | | 351 | | | | | | 220 | | |
| | | | | | | 258 | | | | 323 | | | 326 | | | | 247 | | | | | | |
| | | | | | | | 280 | | | | 300 | 301 | | | 273 | | | | | | | | |
| | | | | | | | 304 | | | | 276 | 277 | | | 297 | | | | | | | | |
| | | | | | | 330 | | | | 251 | | | 254 | | | | 319 | | | | | | |
| | | | 357 | | | | | | 226 | | | | | | 231 | | | | | | 340 | | |
| | | | | | | | 201 | | | | 373 | 372 | | | 208 | | | | | | | | |
| | 406 | | | | | 176 | | | | | | | | | | 185 | | | | | 387 | | |
| | | | | | | 151 | | | | 422 | | | 419 | | | | 162 | | | | | | |
| 456 | | | | | 126 | | | | | | | | | | | | | 139 | | | | | 433 |
| | 479 | | | 101 | | | | | | | | | | | | | | | 116 | | | 458 | |
| | | 76 | | | | | | | 495 | | | | | | 490 | | | | | | 93 | | |
| | 51 | | | | | | 521 | | | | | | | | | 512 | | | | | 70 | | |
| | 26 | | | 548 | | | | | | | | | | | | | | | 533 | | | 47 | |
| 1 | | | | | 571 | | | | | | | | | | | | | 558 | | | | | 24 |

Magic Square – Number Fill-ins [D] – Mathematical Formula 24 X 24 + 1 X 12 = 6924

SHELSEAS MAGIC SQUARE II
It Works

# Solved

# [D]

Mathematical Formulas

8 X 8 + 1 X 4 = 260

12 X 12 + 1 X 6 = 870

16 X 16 + 1 X 8 = 2056

20 X 20 + 1 X 10 = 4010

24 X 24 + 1 X 12 = 6924

Magic Square
Number Fill-ins [D]
Mathematical Formula

# SHELSEAS MAGIC SQUARE II
## It Works

### Positive [D]

|   | 58 | 59 |    |    | 62 | 63 |    |
|---|----|----|----|----|----|----|----|
| 49 | 50 |   |    |    |    | 55 | 56 |
| 41 |   |   | 44 | 45 |    |    | 48 |
|   |   | 35 | 36 | 37 | 38 |    |    |
|   |   | 27 | 28 | 29 | 30 |    |    |
| 17 |   |   | 20 | 21 |    |    | 24 |
| 9 | 10 |   |    |    |    | 15 | 16 |
|   | 2 | 3 |    |    | 6 | 7 |    |

### Negative [D]

| 8 |   |   | 5 | 4 |   |   | 1 |
|---|---|---|---|---|---|---|---|
|   |   | 14 | 13 | 12 | 11 |   |   |
|   | 23 | 22 |   |   | 19 | 18 |   |
| 32 | 31 |   |   |   |   | 26 | 25 |
| 40 | 39 |   |   |   |   | 34 | 33 |
|   | 47 | 46 |   |   | 43 | 42 |   |
|   |   | 54 | 53 | 52 | 51 |   |   |
| 64 |   |   | 61 | 60 |   |   | 57 |

Combination of both Positive & Negative number fill-ins [D] is shown for a solved Magic Square. Objective: Add each square of numbers in any row, column, or diagonal to proof the same sum of 260.

| 8 | 58 | 59 | 5 | 4 | 62 | 63 | 1 |
|---|----|----|---|---|----|----|---|
| 49 | 50 | 14 | 13 | 12 | 11 | 55 | 56 |
| 41 | 23 | 22 | 44 | 45 | 19 | 18 | 48 |
| 32 | 31 | 35 | 36 | 37 | 38 | 26 | 25 |
| 40 | 39 | 27 | 28 | 29 | 30 | 34 | 33 |
| 17 | 47 | 46 | 20 | 21 | 43 | 42 | 24 |
| 9 | 10 | 54 | 53 | 52 | 51 | 15 | 16 |
| 64 | 2 | 3 | 61 | 60 | 6 | 7 | 57 |

Magic Square
Number Fill-ins [D]
Mathematical Formula 8 X 8 + 1 X 4 = 260

# SHELSEAS MAGIC SQUARE II
## It Works

Positive [D] Diagonals are shown for the same sum of 870

| 133 | 134 | 135 |     |     |     |     |     |     | 142 | 143 | 144 |
|-----|-----|-----|-----|-----|-----|-----|-----|-----|-----|-----|-----|
| 121 | 122 |     | 124 |     |     |     |     | 129 |     | 131 | 132 |
| 109 |     | 111 |     | 113 |     |     | 116 |     | 118 |     | 120 |
|     | 98  |     | 100 |     | 102 | 103 |     | 105 |     | 107 |     |
|     |     | 87  |     | 89  | 90  | 91  | 92  |     | 94  |     |     |
|     |     |     | 76  | 77  | 78  | 79  | 80  | 81  |     |     |     |
|     |     |     | 64  | 65  | 66  | 67  | 68  | 69  |     |     |     |
|     |     | 51  |     | 53  | 54  | 55  | 56  |     | 58  |     |     |
|     | 38  |     | 40  |     | 42  | 43  |     | 45  |     | 47  |     |
| 25  |     | 27  |     | 29  |     |     | 32  |     | 34  |     | 36  |
| 13  | 14  |     | 16  |     |     |     |     | 21  |     | 23  | 24  |
| 1   | 2   | 3   |     |     |     |     |     |     | 10  | 11  | 12  |

Negative [D]

|     |     |     | 9   | 8   | 7   | 6   | 5   | 4   |     |     |     |
|-----|-----|-----|-----|-----|-----|-----|-----|-----|-----|-----|-----|
|     |     | 22  |     | 20  | 19  | 18  | 17  |     | 15  |     |     |
|     | 35  |     | 33  |     | 31  | 30  |     | 28  |     | 26  |     |
| 48  |     | 46  |     | 44  |     |     | 41  |     | 39  |     | 37  |
| 60  | 59  |     | 57  |     |     |     |     | 52  |     | 50  | 49  |
| 72  | 71  | 70  |     |     |     |     |     |     | 63  | 62  | 61  |
| 84  | 83  | 82  |     |     |     |     |     |     | 75  | 74  | 73  |
| 96  | 95  |     | 93  |     |     |     |     | 88  |     | 86  | 85  |
| 108 |     | 106 |     | 104 |     |     | 101 |     | 99  |     | 97  |
|     | 119 |     | 117 |     | 115 | 114 |     | 112 |     | 110 |     |
|     |     | 130 |     | 128 | 127 | 126 | 125 |     | 123 |     |     |
|     |     |     | 141 | 140 | 139 | 138 | 137 | 136 |     |     |     |

Combination of both Positive & Negative number fill-ins [D] is shown for a solved Magic Square.  Objective:  Add each square of numbers in any row, column, or diagonal to proof the same sum of 870.

| 133 | 134 | 135 | 9   | 8   | 7   | 6   | 5   | 4   | 142 | 143 | 144 |
|-----|-----|-----|-----|-----|-----|-----|-----|-----|-----|-----|-----|
| 121 | 122 | 22  | 124 | 20  | 19  | 18  | 17  | 129 | 15  | 131 | 132 |
| 109 | 35  | 111 | 33  | 113 | 31  | 30  | 116 | 28  | 118 | 26  | 120 |
| 48  | 98  | 46  | 100 | 44  | 102 | 103 | 41  | 105 | 39  | 107 | 37  |
| 60  | 59  | 87  | 57  | 89  | 90  | 91  | 92  | 52  | 94  | 50  | 49  |
| 72  | 71  | 70  | 76  | 77  | 78  | 79  | 80  | 81  | 63  | 62  | 61  |
| 84  | 83  | 82  | 64  | 65  | 66  | 67  | 68  | 69  | 75  | 74  | 73  |
| 96  | 95  | 51  | 93  | 53  | 54  | 55  | 56  | 88  | 58  | 86  | 85  |
| 108 | 38  | 106 | 40  | 104 | 42  | 43  | 101 | 45  | 99  | 47  | 97  |
| 25  | 119 | 27  | 117 | 29  | 115 | 114 | 32  | 112 | 34  | 110 | 36  |
| 13  | 14  | 130 | 16  | 128 | 127 | 126 | 125 | 21  | 123 | 23  | 24  |
| 1   | 2   | 3   | 141 | 140 | 139 | 138 | 137 | 136 | 10  | 11  | 12  |

Magic Square – Number Fill-ins [D] – Mathematical Formula 12 X 12 + 1 X 6 = 870

# SHELSEAS MAGIC SQUARE II
## It Works

### Positive [D]

|  |  |  |  |  |  |  |  |  |  |  |  |  |  |  |  |
|---|---|---|---|---|---|---|---|---|---|---|---|---|---|---|---|
| 241 | 242 | 243 | 244 |  |  |  |  |  |  |  |  | 253 | 254 | 255 | 256 |
| 225 |  | 227 |  |  |  | 231 | 232 | 233 | 234 |  |  |  | 238 |  | 240 |
| 209 | 210 |  | 212 | 213 |  |  |  |  |  |  | 220 | 221 |  | 223 | 224 |
| 193 |  | 195 |  | 197 | 198 |  |  |  |  | 203 | 204 |  | 206 |  | 208 |
|  |  | 179 | 180 |  | 182 |  | 184 | 185 |  | 187 |  | 189 | 190 |  |  |
|  |  |  | 164 | 165 | 166 | 167 |  |  | 170 | 171 | 172 | 173 |  |  |  |
|  | 146 |  |  |  | 150 | 151 | 152 | 153 | 154 | 155 |  |  |  | 159 |  |
|  | 130 |  |  | 133 |  | 135 | 136 | 137 | 138 |  | 140 |  |  | 143 |  |
|  | 114 |  |  | 117 |  | 119 | 120 | 121 | 122 |  | 124 |  |  | 127 |  |
|  | 98 |  |  |  | 102 | 103 | 104 | 105 | 106 | 107 |  |  |  | 111 |  |
|  |  |  | 84 | 85 | 86 | 87 |  |  | 90 | 91 | 92 | 93 |  |  |  |
|  |  | 67 | 68 |  | 70 |  | 72 | 73 |  | 75 |  | 77 | 78 |  |  |
| 49 |  | 51 |  | 53 | 54 |  |  |  |  | 59 | 60 |  | 62 |  | 64 |
| 33 | 34 |  | 36 | 37 |  |  |  |  |  |  | 44 | 45 |  | 47 | 48 |
| 17 |  | 19 |  |  |  | 23 | 24 | 25 | 26 |  |  |  | 30 |  | 32 |
| 1 | 2 | 3 | 4 |  |  |  |  |  |  |  |  | 13 | 14 | 15 | 16 |

### Negative [D]

|  |  |  |  |  |  |  |  |  |  |  |  |  |  |  |  |
|---|---|---|---|---|---|---|---|---|---|---|---|---|---|---|---|
|  |  |  |  | 12 | 11 | 10 | 9 | 8 | 7 | 6 | 5 |  |  |  |  |
|  | 31 |  | 29 | 28 | 27 |  |  |  |  | 22 | 21 | 20 |  | 18 |  |
|  |  | 46 |  |  | 43 | 42 | 41 | 40 | 39 | 38 |  |  | 35 |  |  |
|  | 63 |  | 61 |  |  | 58 | 57 | 56 | 55 |  |  | 52 |  | 50 |  |
| 80 | 79 |  |  | 76 |  | 74 |  |  | 71 |  | 69 |  |  | 66 | 65 |
| 96 | 95 | 94 |  |  |  |  | 89 | 88 |  |  |  |  | 83 | 82 | 81 |
| 112 |  | 110 | 109 | 108 |  |  |  |  |  |  | 101 | 100 | 99 |  | 97 |
| 128 |  | 126 | 125 |  | 123 |  |  |  |  | 118 |  | 116 | 115 |  | 113 |
| 144 |  | 142 | 141 |  | 139 |  |  |  |  | 134 |  | 132 | 131 |  | 129 |
| 160 |  | 158 | 157 | 156 |  |  |  |  |  |  | 149 | 148 | 147 |  | 145 |
| 176 | 175 | 174 |  |  |  |  | 169 | 168 |  |  |  |  | 163 | 162 | 161 |
| 192 | 191 |  |  | 188 |  | 186 |  |  | 183 |  | 181 |  |  | 178 | 177 |
|  | 207 |  | 205 |  |  | 202 | 201 | 200 | 199 |  |  | 196 |  | 194 |  |
|  |  | 222 |  |  | 219 | 218 | 217 | 216 | 215 | 214 |  |  | 211 |  |  |
|  | 239 |  | 237 | 236 | 235 |  |  |  |  | 230 | 229 | 228 |  | 226 |  |
|  |  |  |  | 252 | 251 | 250 | 249 | 248 | 247 | 246 | 245 |  |  |  |  |

Magic Square
Number Fill-ins [D]
Mathematical Formula 16 X 16 + 1 X 8 = 2056

# SHELSEAS MAGIC SQUARE II
## It Works

Combination of both Positive & Negative number fill-ins [D] is shown for a solved Magic Square. Objective: Add each square of numbers in any row, column, or diagonal to proof the same sum of 2056.

| 241 | 242 | 243 | 244 | 12 | 11 | 10 | 9 | 8 | 7 | 6 | 5 | 253 | 254 | 255 | 256 |
|-----|-----|-----|-----|-----|-----|-----|-----|-----|-----|-----|-----|-----|-----|-----|-----|
| 225 | 31 | 227 | 29 | 28 | 27 | 231 | 232 | 233 | 234 | 22 | 21 | 20 | 238 | 18 | 240 |
| 209 | 210 | 46 | 212 | 213 | 43 | 42 | 41 | 40 | 39 | 38 | 220 | 221 | 35 | 223 | 224 |
| 193 | 63 | 195 | 61 | 197 | 198 | 58 | 57 | 56 | 55 | 203 | 204 | 52 | 206 | 50 | 208 |
| 80 | 79 | 179 | 180 | 76 | 182 | 74 | 184 | 185 | 71 | 187 | 69 | 189 | 190 | 66 | 65 |
| 96 | 95 | 94 | 164 | 165 | 166 | 167 | 89 | 88 | 170 | 171 | 172 | 173 | 83 | 82 | 81 |
| 112 | 146 | 110 | 109 | 108 | 150 | 151 | 152 | 153 | 154 | 155 | 101 | 100 | 99 | 159 | 97 |
| 128 | 130 | 126 | 125 | 133 | 123 | 135 | 136 | 137 | 138 | 118 | 140 | 116 | 115 | 143 | 113 |
| 144 | 114 | 142 | 141 | 117 | 139 | 119 | 120 | 121 | 122 | 134 | 124 | 132 | 131 | 127 | 129 |
| 160 | 98 | 158 | 157 | 156 | 102 | 103 | 104 | 105 | 106 | 107 | 149 | 148 | 147 | 111 | 145 |
| 176 | 175 | 174 | 84 | 85 | 86 | 87 | 169 | 168 | 90 | 91 | 92 | 93 | 163 | 162 | 161 |
| 192 | 191 | 67 | 68 | 188 | 70 | 186 | 72 | 73 | 183 | 75 | 181 | 77 | 78 | 178 | 177 |
| 49 | 207 | 51 | 205 | 53 | 54 | 202 | 201 | 200 | 199 | 59 | 60 | 196 | 62 | 194 | 64 |
| 33 | 34 | 222 | 36 | 37 | 219 | 218 | 217 | 216 | 215 | 214 | 44 | 45 | 211 | 47 | 48 |
| 17 | 239 | 19 | 237 | 236 | 235 | 23 | 24 | 25 | 26 | 230 | 229 | 228 | 30 | 226 | 32 |
| 1 | 2 | 3 | 4 | 252 | 251 | 250 | 249 | 248 | 247 | 246 | 245 | 13 | 14 | 15 | 16 |

# 2056

Magic Square
Number Fill-ins [D]
Mathematical Formula 16 X 16 + 1 X 8 = 2056

# SHELSEAS MAGIC SQUARE II
## It Works

### Positive [D]

| | | | | | | | | | | | | | | | | | | | |
|---|---|---|---|---|---|---|---|---|---|---|---|---|---|---|---|---|---|---|---|
| 381 | 382 | 383 | 384 | | | | | | 390 | 391 | | | | | | 397 | 398 | 399 | 400 |
| 361 | 362 | 363 | 364 | 365 | | | | | | | | | | | 376 | 377 | 378 | 379 | 380 |
| 341 | 342 | | | 345 | 346 | 347 | | | | | | | 354 | 355 | 356 | | | 359 | 360 |
| 321 | 322 | | | | 326 | 327 | 328 | | | | | 333 | 334 | 335 | | | | 339 | 340 |
| | 302 | 303 | | | | 307 | 308 | 309 | | | 312 | 313 | 314 | | | | 318 | 319 | |
| | | 283 | 284 | | | | 288 | 289 | 290 | 291 | 292 | 293 | | | | 297 | 298 | | |
| | | 263 | 264 | 265 | | | | 269 | 270 | 271 | 272 | | | | 276 | 277 | 278 | | |
| | | | 244 | 245 | 246 | | | 249 | 250 | 251 | 252 | | | 255 | 256 | 257 | | | |
| | | | | 225 | 226 | 227 | 228 | 229 | | | 232 | 233 | 234 | 235 | 236 | | | | |
| 201 | | | | | 206 | 207 | 208 | | 210 | 211 | | 213 | 214 | 215 | | | | | 220 |
| 181 | | | | | 186 | 187 | 188 | | 190 | 191 | | 193 | 194 | 195 | | | | | 200 |
| | | | | 165 | 166 | 167 | 168 | 169 | | | 172 | 173 | 174 | 175 | 176 | | | | |
| | | | 144 | 145 | 146 | | | 149 | 150 | 151 | 152 | | | 155 | 156 | 157 | | | |
| | | 123 | 124 | 125 | | | | 129 | 130 | 131 | 132 | | | | 136 | 137 | 138 | | |
| | | 103 | 104 | | | | 108 | 109 | 110 | 111 | 112 | 113 | | | | 117 | 118 | | |
| | 82 | 83 | | | | 87 | 88 | 89 | | | 92 | 93 | 94 | | | | 98 | 99 | |
| 61 | 62 | | | | 66 | 67 | 68 | | | | | 73 | 74 | 75 | | | | 79 | 80 |
| 41 | 42 | | | 45 | 46 | 47 | | | | | | | 54 | 55 | 56 | | | 59 | 60 |
| 21 | 22 | 23 | 24 | 25 | | | | | | | | | | | 36 | 37 | 38 | 39 | 40 |
| 1 | 2 | 3 | 4 | | | | | | 10 | 11 | | | | | | 17 | 18 | 19 | 20 |

### Negative [D]

| | | | | | | | | | | | | | | | | | | | |
|---|---|---|---|---|---|---|---|---|---|---|---|---|---|---|---|---|---|---|---|
| | | | | 16 | 15 | 14 | 13 | 12 | | | 9 | 8 | 7 | 6 | 5 | | | | |
| | | | | | 35 | 34 | 33 | 32 | 31 | 30 | 29 | 28 | 27 | 26 | | | | | |
| | | 58 | 57 | | | | 53 | 52 | 51 | 50 | 49 | 48 | | | | 44 | 43 | | |
| | | 78 | 77 | 76 | | | | 72 | 71 | 70 | 69 | | | | 65 | 64 | 63 | | |
| 100 | | | 97 | 96 | 95 | | | | 91 | 90 | | | | 86 | 85 | 84 | | | 81 |
| 120 | 119 | | | 116 | 115 | 114 | | | | | | | 107 | 106 | 105 | | | 102 | 101 |
| 140 | 139 | | | | 135 | 134 | 133 | | | | | 128 | 127 | 126 | | | | 122 | 121 |
| 160 | 159 | 158 | | | | 154 | 153 | | | | | 148 | 147 | | | | 143 | 142 | 141 |
| 180 | 179 | 178 | 177 | | | | | | 171 | 170 | | | | | | 164 | 163 | 162 | 161 |
| | 199 | 198 | 197 | 196 | | | | 192 | | | 189 | | | | 185 | 184 | 183 | 182 | |
| | 219 | 218 | 217 | 216 | | | | 212 | | | 209 | | | | 205 | 204 | 203 | 202 | |
| 240 | 239 | 238 | 237 | | | | | | 231 | 230 | | | | | | 224 | 223 | 222 | 221 |
| 260 | 259 | 258 | | | | 254 | 253 | | | | | 248 | 247 | | | | 243 | 242 | 241 |
| 280 | 279 | | | | 275 | 274 | 273 | | | | | 268 | 267 | 266 | | | | 262 | 261 |
| 300 | 299 | | | 296 | 295 | 294 | | | | | | | 287 | 286 | 285 | | | 282 | 281 |
| 320 | | | 317 | 316 | 315 | | | | 311 | 310 | | | | 306 | 305 | 304 | | | 301 |
| | | 338 | 337 | 336 | | | | 332 | 331 | 330 | 329 | | | | 325 | 324 | 323 | | |
| | | 358 | 357 | | | | 353 | 352 | 351 | 350 | 349 | 348 | | | | 344 | 343 | | |
| | | | | | 375 | 374 | 373 | 372 | 371 | 370 | 369 | 368 | 367 | 366 | | | | | |
| | | | | 396 | 395 | 394 | 393 | 392 | | | 389 | 388 | 387 | 386 | 385 | | | | |

Magic Square – Number Fill-ins [D] – Mathematical Formula 20 X 20 + 1 X 10 = 4010

Combination of both Positive & Negative number fill-ins [D] is shown for a solved Magic Square.  Objective:  Add each square of numbers in any row, column, or diagonal to proof the same sum of 4010.

| 381 | 382 | 383 | 384 | 16 | 15 | 14 | 13 | 12 | 390 | 391 | 9 | 8 | 7 | 6 | 5 | 397 | 398 | 399 | 400 |
|-----|-----|-----|-----|-----|-----|-----|-----|-----|-----|-----|-----|-----|-----|-----|-----|-----|-----|-----|-----|
| 361 | 362 | 363 | 364 | 365 | 35 | 34 | 33 | 32 | 31 | 30 | 29 | 28 | 27 | 26 | 376 | 377 | 378 | 379 | 380 |
| 341 | 342 | 58 | 57 | 345 | 346 | 347 | 53 | 52 | 51 | 50 | 49 | 48 | 354 | 355 | 356 | 44 | 43 | 359 | 360 |
| 321 | 322 | 78 | 77 | 76 | 326 | 327 | 328 | 72 | 71 | 70 | 69 | 333 | 334 | 335 | 65 | 64 | 63 | 339 | 340 |
| 100 | 302 | 303 | 97 | 96 | 95 | 307 | 308 | 309 | 91 | 90 | 312 | 313 | 314 | 86 | 85 | 84 | 318 | 319 | 81 |
| 120 | 119 | 283 | 284 | 116 | 115 | 114 | 288 | 289 | 290 | 291 | 292 | 293 | 107 | 106 | 105 | 297 | 298 | 102 | 101 |
| 140 | 139 | 263 | 264 | 265 | 135 | 134 | 133 | 269 | 270 | 271 | 272 | 128 | 127 | 126 | 276 | 277 | 278 | 122 | 121 |
| 160 | 159 | 158 | 244 | 245 | 246 | 154 | 153 | 249 | 250 | 251 | 252 | 148 | 147 | 255 | 256 | 257 | 143 | 142 | 141 |
| 180 | 179 | 178 | 177 | 225 | 226 | 227 | 228 | 229 | 171 | 170 | 232 | 233 | 234 | 235 | 236 | 164 | 163 | 162 | 161 |
| 201 | 199 | 198 | 197 | 196 | 206 | 207 | 208 | 192 | 210 | 211 | 189 | 213 | 214 | 215 | 185 | 184 | 183 | 182 | 220 |
| 181 | 219 | 218 | 217 | 216 | 186 | 187 | 188 | 212 | 190 | 191 | 209 | 193 | 194 | 195 | 205 | 204 | 203 | 202 | 200 |
| 240 | 239 | 238 | 237 | 165 | 166 | 167 | 168 | 169 | 231 | 230 | 172 | 173 | 174 | 175 | 176 | 224 | 223 | 222 | 221 |
| 260 | 259 | 258 | 144 | 145 | 146 | 254 | 253 | 149 | 150 | 151 | 152 | 248 | 247 | 155 | 156 | 157 | 243 | 242 | 241 |
| 280 | 279 | 123 | 124 | 125 | 275 | 274 | 273 | 129 | 130 | 131 | 132 | 268 | 267 | 266 | 136 | 137 | 138 | 262 | 261 |
| 300 | 299 | 103 | 104 | 296 | 295 | 294 | 108 | 109 | 110 | 111 | 112 | 113 | 287 | 286 | 285 | 117 | 118 | 282 | 281 |
| 320 | 82 | 83 | 317 | 316 | 315 | 87 | 88 | 89 | 311 | 310 | 92 | 93 | 94 | 306 | 305 | 304 | 98 | 99 | 301 |
| 61 | 62 | 338 | 337 | 336 | 66 | 67 | 68 | 332 | 331 | 330 | 329 | 73 | 74 | 75 | 325 | 324 | 323 | 79 | 80 |
| 41 | 42 | 358 | 357 | 45 | 46 | 47 | 353 | 352 | 351 | 350 | 349 | 348 | 54 | 55 | 56 | 344 | 343 | 59 | 60 |
| 21 | 22 | 23 | 24 | 25 | 375 | 374 | 373 | 372 | 371 | 370 | 369 | 368 | 367 | 366 | 36 | 37 | 38 | 39 | 40 |
| 1 | 2 | 3 | 4 | 396 | 395 | 394 | 393 | 392 | 10 | 11 | 389 | 388 | 387 | 386 | 385 | 17 | 18 | 19 | 20 |

# 4010

Magic Square
Number Fill-ins [D]
Mathematical Formula 20 X 20 + 1 X 10 = 4010

# SHELSEAS MAGIC SQUARE II
## It Works

### Positive [D] Diagonals are shown for the same sum of 6924

|  |  |  |  |  |  |  |  |  |  |  |  |  |  |  |  |  |  |  |  |  |  |  |  |
|---|---|---|---|---|---|---|---|---|---|---|---|---|---|---|---|---|---|---|---|---|---|---|---|
| 553 | 554 | 555 | 556 | 557 |  |  |  |  |  |  | 564 | 565 |  |  |  |  |  |  | 572 | 573 | 574 | 575 | 576 |
| 529 | 530 | 531 | 532 |  |  |  |  |  |  | 539 | 540 | 541 | 542 |  |  |  |  |  |  | 549 | 550 | 551 | 552 |
| 505 | 506 | 507 | 508 | 509 |  |  |  |  |  |  | 516 | 517 |  |  |  |  |  |  | 524 | 525 | 526 | 527 | 528 |
| 481 | 482 | 483 | 484 | 485 | 486 |  |  |  |  |  |  |  |  |  |  |  |  | 499 | 500 | 501 | 502 | 503 | 504 |
| 457 |  | 459 | 460 | 461 | 462 | 463 |  |  |  |  |  |  |  |  |  |  | 474 | 475 | 476 | 477 | 478 |  | 480 |
|  |  |  | 436 | 437 | 438 | 439 | 440 | 441 |  |  |  |  |  |  | 448 | 449 | 450 | 451 | 452 | 453 |  |  |  |
|  |  |  |  | 413 | 414 | 415 | 416 | 417 | 418 |  |  |  |  | 423 | 424 | 425 | 426 | 427 | 428 |  |  |  |  |
|  |  |  |  |  | 390 | 391 | 392 | 393 | 394 | 395 |  |  | 398 | 399 | 400 | 401 | 402 | 403 |  |  |  |  |  |
|  |  |  |  |  | 366 | 367 | 368 | 369 | 370 | 371 |  |  | 374 | 375 | 376 | 377 | 378 | 379 |  |  |  |  |  |
|  |  |  |  |  |  | 343 | 344 | 345 | 346 | 347 | 348 | 349 | 350 | 351 | 352 | 353 | 354 |  |  |  |  |  |  |
|  | 314 |  |  |  |  |  | 320 | 321 | 322 | 323 | 324 | 325 | 326 | 327 | 328 | 329 |  |  |  |  |  | 335 |  |
| 289 | 290 | 291 |  |  |  |  |  |  | 298 | 299 | 300 | 301 | 302 | 303 |  |  |  |  |  |  | 310 | 311 | 312 |
| 265 | 266 | 267 |  |  |  |  |  |  | 274 | 275 | 276 | 277 | 278 | 279 |  |  |  |  |  |  | 286 | 287 | 288 |
|  | 242 |  |  |  |  |  | 248 | 249 | 250 | 251 | 252 | 253 | 254 | 255 | 256 | 257 |  |  |  |  |  | 263 |  |
|  |  |  |  |  |  | 223 | 224 | 225 | 226 | 227 | 228 | 229 | 230 | 231 | 232 | 233 | 234 |  |  |  |  |  |  |
|  |  |  |  |  | 198 | 199 | 200 | 201 | 202 | 203 |  |  | 206 | 207 | 208 | 209 | 210 | 211 |  |  |  |  |  |
|  |  |  |  |  | 174 | 175 | 176 | 177 | 178 | 179 |  |  | 182 | 183 | 184 | 185 | 186 | 187 |  |  |  |  |  |
|  |  |  |  | 149 | 150 | 151 | 152 | 153 | 154 |  |  |  |  | 159 | 160 | 161 | 162 | 163 | 164 |  |  |  |  |
|  |  |  | 124 | 125 | 126 | 127 | 128 | 129 |  |  |  |  |  |  | 136 | 137 | 138 | 139 | 140 | 141 |  |  |  |
| 97 |  | 99 | 100 | 101 | 102 | 103 |  |  |  |  |  |  |  |  |  |  | 114 | 115 | 116 | 117 | 118 |  | 120 |
| 73 | 74 | 75 | 76 | 77 | 78 |  |  |  |  |  |  |  |  |  |  |  |  | 91 | 92 | 93 | 94 | 95 | 96 |
| 49 | 50 | 51 | 52 | 53 |  |  |  |  |  |  | 60 | 61 |  |  |  |  |  |  | 68 | 69 | 70 | 71 | 72 |
| 25 | 26 | 27 | 28 |  |  |  |  |  |  | 35 | 36 | 37 | 38 |  |  |  |  |  |  | 45 | 46 | 47 | 48 |
| 1 | 2 | 3 | 4 | 5 |  |  |  |  |  |  | 12 | 13 |  |  |  |  |  |  | 20 | 21 | 22 | 23 | 24 |

### Negative [D]

|  |  |  |  |  |  |  |  |  |  |  |  |  |  |  |  |  |  |  |  |  |  |  |  |
|---|---|---|---|---|---|---|---|---|---|---|---|---|---|---|---|---|---|---|---|---|---|---|---|
|  |  |  |  |  | 19 | 18 | 17 | 16 | 15 | 14 |  |  | 11 | 10 | 9 | 8 | 7 | 6 |  |  |  |  |  |
|  |  |  |  | 44 | 43 | 42 | 41 | 40 | 39 |  |  |  |  | 34 | 33 | 32 | 31 | 30 | 29 |  |  |  |  |
|  |  |  |  |  | 67 | 66 | 65 | 64 | 63 | 62 |  |  | 59 | 58 | 57 | 56 | 55 | 54 |  |  |  |  |  |
|  |  |  |  |  |  | 90 | 89 | 88 | 87 | 86 | 85 | 84 | 83 | 82 | 81 | 80 | 79 |  |  |  |  |  |  |
|  | 119 |  |  |  |  |  | 113 | 112 | 111 | 110 | 109 | 108 | 107 | 106 | 105 | 104 |  |  |  |  |  | 98 |  |
| 144 | 143 | 142 |  |  |  |  |  |  | 135 | 134 | 133 | 132 | 131 | 130 |  |  |  |  |  |  | 123 | 122 | 121 |
| 168 | 167 | 166 | 165 |  |  |  |  |  |  | 158 | 157 | 156 | 155 |  |  |  |  |  |  | 148 | 147 | 146 | 145 |
| 192 | 191 | 190 | 189 | 188 |  |  |  |  |  |  | 181 | 180 |  |  |  |  |  |  | 173 | 172 | 171 | 170 | 169 |
| 216 | 215 | 214 | 213 | 212 |  |  |  |  |  |  | 205 | 204 |  |  |  |  |  |  | 197 | 196 | 195 | 194 | 193 |
| 240 | 239 | 238 | 237 | 236 | 235 |  |  |  |  |  |  |  |  |  |  |  |  | 222 | 221 | 220 | 219 | 218 | 217 |
| 264 |  | 262 | 261 | 260 | 259 | 258 |  |  |  |  |  |  |  |  |  |  | 247 | 246 | 245 | 244 | 243 |  | 241 |
|  |  |  | 285 | 284 | 283 | 282 | 281 | 280 |  |  |  |  |  |  | 273 | 272 | 271 | 270 | 269 | 268 |  |  |  |
|  |  |  | 309 | 308 | 307 | 306 | 305 | 304 |  |  |  |  |  |  | 297 | 296 | 295 | 294 | 293 | 292 |  |  |  |
| 336 |  | 334 | 333 | 332 | 331 | 330 |  |  |  |  |  |  |  |  |  |  | 319 | 318 | 317 | 316 | 315 |  | 313 |
| 360 | 359 | 358 | 357 | 356 | 355 |  |  |  |  |  |  |  |  |  |  |  |  | 342 | 341 | 340 | 339 | 338 | 337 |
| 384 | 383 | 382 | 381 | 380 |  |  |  |  |  |  | 373 | 372 |  |  |  |  |  |  | 365 | 364 | 363 | 362 | 361 |
| 408 | 407 | 406 | 405 | 404 |  |  |  |  |  |  | 397 | 396 |  |  |  |  |  |  | 389 | 388 | 387 | 386 | 385 |
| 432 | 431 | 430 | 429 |  |  |  |  |  |  | 422 | 421 | 420 | 419 |  |  |  |  |  |  | 412 | 411 | 410 | 409 |
| 456 | 455 | 454 |  |  |  |  |  |  | 447 | 446 | 445 | 444 | 443 | 442 |  |  |  |  |  |  | 435 | 434 | 433 |
|  | 479 |  |  |  |  |  | 473 | 472 | 471 | 470 | 469 | 468 | 467 | 466 | 465 | 464 |  |  |  |  |  | 458 |  |
|  |  |  |  |  |  | 498 | 497 | 496 | 495 | 494 | 493 | 492 | 491 | 490 | 489 | 488 | 487 |  |  |  |  |  |  |
|  |  |  |  |  | 523 | 522 | 521 | 520 | 519 | 518 |  |  | 515 | 514 | 513 | 512 | 511 | 510 |  |  |  |  |  |
|  |  |  |  | 548 | 547 | 546 | 545 | 544 | 543 |  |  |  |  | 538 | 537 | 536 | 535 | 534 | 533 |  |  |  |  |
|  |  |  |  |  | 571 | 570 | 569 | 568 | 567 | 566 |  |  | 563 | 562 | 561 | 560 | 559 | 558 |  |  |  |  |  |

Magic Square – Number Fill-ins [D] – Mathematical Formula 24 X 24 + 1 X 12 = 6924

# SHELSEAS MAGIC SQUARE II
## It Works

Combination of both Positive & Negative number fill-ins [D] is shown for a solved Magic Square.  Objective:  Add each square of numbers in any row, column, or diagonal to proof the same sum of 6924.

| | | | | | | | | | | | | | | | | | | | | | | | |
|---|---|---|---|---|---|---|---|---|---|---|---|---|---|---|---|---|---|---|---|---|---|---|---|
| 553 | 554 | 555 | 556 | 557 | 19 | 18 | 17 | 16 | 15 | 14 | 564 | 565 | 11 | 10 | 9 | 8 | 7 | 6 | 572 | 573 | 574 | 575 | 576 |
| 529 | 530 | 531 | 532 | 44 | 43 | 42 | 41 | 40 | 39 | 539 | 540 | 541 | 542 | 34 | 33 | 32 | 31 | 30 | 29 | 549 | 550 | 551 | 552 |
| 505 | 506 | 507 | 508 | 509 | 67 | 66 | 65 | 64 | 63 | 62 | 516 | 517 | 59 | 58 | 57 | 56 | 55 | 54 | 524 | 525 | 526 | 527 | 528 |
| 481 | 482 | 483 | 484 | 485 | 486 | 90 | 89 | 88 | 87 | 86 | 85 | 84 | 83 | 82 | 81 | 80 | 79 | 499 | 500 | 501 | 502 | 503 | 504 |
| 457 | 119 | 459 | 460 | 461 | 462 | 463 | 113 | 112 | 111 | 110 | 109 | 108 | 107 | 106 | 105 | 104 | 474 | 475 | 476 | 477 | 478 | 98 | 480 |
| 144 | 143 | 142 | 436 | 437 | 438 | 439 | 440 | 441 | 135 | 134 | 133 | 132 | 131 | 130 | 448 | 449 | 450 | 451 | 452 | 453 | 123 | 122 | 121 |
| 168 | 167 | 166 | 165 | 413 | 414 | 415 | 416 | 417 | 418 | 158 | 157 | 156 | 155 | 423 | 424 | 425 | 426 | 427 | 428 | 148 | 147 | 146 | 145 |
| 192 | 191 | 190 | 189 | 188 | 390 | 391 | 392 | 393 | 394 | 395 | 181 | 180 | 398 | 399 | 400 | 401 | 402 | 403 | 173 | 172 | 171 | 170 | 169 |
| 216 | 215 | 214 | 213 | 212 | 366 | 367 | 368 | 369 | 370 | 371 | 205 | 204 | 374 | 375 | 376 | 377 | 378 | 379 | 197 | 196 | 195 | 194 | 193 |
| 240 | 239 | 238 | 237 | 236 | 235 | 343 | 344 | 345 | 346 | 347 | 348 | 349 | 350 | 351 | 352 | 353 | 354 | 222 | 221 | 220 | 219 | 218 | 217 |
| 264 | 314 | 262 | 261 | 260 | 259 | 258 | 320 | 321 | 322 | 323 | 324 | 325 | 326 | 327 | 328 | 329 | 247 | 246 | 245 | 244 | 243 | 335 | 241 |
| 289 | 290 | 291 | 285 | 284 | 283 | 282 | 281 | 280 | 298 | 299 | 300 | 301 | 302 | 303 | 273 | 272 | 271 | 270 | 269 | 268 | 310 | 311 | 312 |
| 265 | 266 | 267 | 309 | 308 | 307 | 306 | 305 | 304 | 274 | 275 | 276 | 277 | 278 | 279 | 297 | 296 | 295 | 294 | 293 | 292 | 286 | 287 | 288 |
| 336 | 242 | 334 | 333 | 332 | 331 | 330 | 248 | 249 | 250 | 251 | 252 | 253 | 254 | 255 | 256 | 257 | 319 | 318 | 317 | 316 | 315 | 263 | 313 |
| 360 | 359 | 358 | 357 | 356 | 355 | 223 | 224 | 225 | 226 | 227 | 228 | 229 | 230 | 231 | 232 | 233 | 234 | 342 | 341 | 340 | 339 | 338 | 337 |
| 384 | 383 | 382 | 381 | 380 | 198 | 199 | 200 | 201 | 202 | 203 | 373 | 372 | 206 | 207 | 208 | 209 | 210 | 211 | 365 | 364 | 363 | 362 | 361 |
| 408 | 407 | 406 | 405 | 404 | 174 | 175 | 176 | 177 | 178 | 179 | 397 | 396 | 182 | 183 | 184 | 185 | 186 | 187 | 389 | 388 | 387 | 386 | 385 |
| 432 | 431 | 430 | 429 | 149 | 150 | 151 | 152 | 153 | 154 | 422 | 421 | 420 | 419 | 159 | 160 | 161 | 162 | 163 | 164 | 412 | 411 | 410 | 409 |
| 456 | 455 | 454 | 124 | 125 | 126 | 127 | 128 | 129 | 447 | 446 | 445 | 444 | 443 | 442 | 136 | 137 | 138 | 139 | 140 | 141 | 435 | 434 | 433 |
| 97 | 479 | 99 | 100 | 101 | 102 | 103 | 473 | 472 | 471 | 470 | 469 | 468 | 467 | 466 | 465 | 464 | 114 | 115 | 116 | 117 | 118 | 458 | 120 |
| 73 | 74 | 75 | 76 | 77 | 78 | 498 | 497 | 496 | 495 | 494 | 493 | 492 | 491 | 490 | 489 | 488 | 487 | 91 | 92 | 93 | 94 | 95 | 96 |
| 49 | 50 | 51 | 52 | 53 | 523 | 522 | 521 | 520 | 519 | 518 | 60 | 61 | 515 | 514 | 513 | 512 | 511 | 510 | 68 | 69 | 70 | 71 | 72 |
| 25 | 26 | 27 | 28 | 548 | 547 | 546 | 545 | 544 | 543 | 35 | 36 | 37 | 38 | 538 | 537 | 536 | 535 | 534 | 533 | 45 | 46 | 47 | 48 |
| 1 | 2 | 3 | 4 | 5 | 571 | 570 | 569 | 568 | 567 | 566 | 12 | 13 | 563 | 562 | 561 | 560 | 559 | 558 | 20 | 21 | 22 | 23 | 24 |

# 6924

Magic Square
Number Fill-ins [D]
Mathematical Formula 24 X 24 + 1 X 12 = 6924